GRAND CENTRAL'S ENGINEER

THE JOHNS HOPKINS UNIVERSITY STUDIES IN
HISTORICAL AND POLITICAL SCIENCE
130th Series (2012)

1. Kurt C. Schlichting, *Grand Central's Engineer:
William J. Wilgus and the Planning of Modern Manhattan*

2. Joseph November, *Biomedical Computing:
Digitizing Life in the United States*

3. Robert Fox, *The Savant and the State:
Science and Cultural Politics in Nineteenth-Century France*

GRAND CENTRAL'S ENGINEER

WILLIAM J. WILGUS AND THE
PLANNING OF MODERN MANHATTAN

KURT C. SCHLICHTING

The Johns Hopkins University Press
BALTIMORE

© 2012 The Johns Hopkins University Press
All rights reserved. Published 2012
Printed in the United States of America on acid-free paper

Johns Hopkins Paperback edition, 2013
9 8 7 6 5 4 3 2 1

The Johns Hopkins University Press
2715 North Charles Street
Baltimore, Maryland 21218-4363
www.press.jhu.edu

The Library of Congress has cataloged the hardcover edition of this book as follows:
Schlichting, Kurt C.
    Grand Central's engineer : William J. Wilgus and the planning of modern
Manhattan / Kurt C. Schlichting.
        p.   cm. — (The Johns Hopkins University studies in historical and political
science)
    Includes bibliographical references and index.
    ISBN-13: 978-1-4214-0302-1 (hardcover : alk. paper)
    ISBN-10: 1-4214-0302-1 (hardcover : alk. paper)
    1. Wilgus, William J. (William John), 1865–1949.   2. Civil engineers—United
States—Biography.   3. Grand Central Terminal (New York, N.Y.)   4. Manhattan
(New York, N.Y.)—Buildings, structures, etc.—History—20th century.   5. Transportation engineering—New York Metropolitan Area—History—20th century.   I. Title.
    TA140.W54S35   2012
    624.092—dc23
    [B]   2011019913

A catalog record for this book is available from the British Library.

    ISBN-13: 978-1-4214-1193-4
    ISBN-10: 1-4214-1193-8

Frontispiece: William J. Wilgus, circa 1900

*Special discounts are available for bulk purchases of this book. For more information,
please contact Special Sales at 410-516-6936 or specialsales@press.jhu.edu.*

The Johns Hopkins University Press uses environmentally friendly book materials,
including recycled text paper that is composed of at least 30 percent post-consumer
waste, whenever possible.

The last printed page of the book constitutes an extension of this copyright page.

To

DR. ARTHUR ANDERSON, FAIRFIELD UNIVERSITY

DR. RICHARD MAISEL, NEW YORK UNIVERSITY

We all need good mentors and wise counsel—Thank you.

Ah, what can ever be more stately and admirable to me
than mast-hemm'd Manhattan?

—WALT WHITMAN, *Crossing Brooklyn Ferry*

CONTENTS

PREFACE

WILLIAM J. WILGUS left a mark on the City of New York. His masterpiece, Grand Central Terminal and the underground train yard, forever changed midtown Manhattan and created a "Terminal City" surrounding the new facility opened in February 1913. Soon all of New York and the country will celebrate Grand Central's 100th anniversary. When Wilgus first presented his ideas for a new Grand Central to the senior management of the New York Central & Hudson River Railroad in 1902, he was just thirty-seven years old. Recently promoted by the Central to the position of fifth vice president and chief engineer, he boldly convinced the railroad to spend a fortune on its flagship terminal in the heart of New York City.

Wilgus recognized the key challenges created by the geography of New York City. Manhattan is a narrow island surrounded by rivers, with the greatest harbor in the world to the south. New York rose to prominence because of its harbor and connections by water to the world. Millions of immigrants to the United States came by ship through New York. Those who stayed provided the labor for the city's industrial and commercial ascendency and turned Manhattan into the most densely populated island in the world at the time. Wilgus developed a series of innovative engineering projects designed to facilitate the movement of people and goods to and from Manhattan and throughout the harbor and region.

The primary sources for my book on Grand Central were the William Wilgus Papers in the Manuscripts Division of the New York Public Library. His vast collection of materials also documents a long career that followed his triumph at Grand Central, a career that prospered after the Grand Central project and engaged his talent and imagination in a relentless effort to improve transportation to Manhattan and the port of New York.

Above all else, William Wilgus remained steadfast in his belief that the railroad and mass transit offered the only viable transportation in the City of New York and the surrounding metropolitan region. When he arrived in New York at the turn of the twentieth century, the age of the automobile and truck were barely under way; the railroads and mass transportation ruled supreme. Later in his career he worked as the chief consulting engineer on the construction of the Holland Tunnel, the first vehicular tunnel under the Hudson River linking Manhattan to New Jersey. Even while working on the Holland Tunnel, he continued to advocate for improved rail transportation.

Wilgus recognized that the transportation challenges in New York were both geographical and political. Powerful railroads and shipping interests resisted his efforts to bring a rational planning model to bear on the city's transport needs. Local politicians resisted cooperating in his plans to efficiently link the New York side of the Hudson River and New York harbor with New Jersey. To overcome parochial public and private interests in the port, he proposed the creation of a bi-state agency that would have authority over the entire port and region and bring order to the transportation chaos. From Wilgus's fertile mind came the idea for the Port Authority of New York and New Jersey.

A BOOK BEGINS with an idea, but that idea needs encouragement. After the publication of *Grand Central Terminal: Railroads, Engineering, and Architecture in New York City* by the Johns Hopkins University Press, my editor, Bob Brugger, asked whether a book on William Wilgus's career, focused on his later work in New York, made sense. More conversations led to a proposal for this book. Bob's encouragement was crucial. Carolyn Moser did a superb job editing the manuscript, and Bill Nelson, a gifted cartographer, created the maps. I also thank Kara Reiter for her assistance in keeping track of the myriad details of maps, photographs, and permissions and Linda Forlifer for overseeing the publication of the book.

Once again the Manuscript Division of the New York Public Library provided unparalleled assistance with access to the Wilgus papers. The staff of the Manuscript Division are superb. Melanie Yolles, Senior Archivist, led me to the Oppenheim papers, which include letters from Wilgus's son about his father, a very private man. Thomas Lamont, Reference Archivist, assisted with photographs from the Wilgus collection and others from the collection of the library.

As incumbent E. Gerald Corrigan '63 Chair in Humanities and Social Sciences at Fairfield University, I offer special thanks to E. Gerald Corrigan, Ph.D., Class of '63, for his generous support to Fairfield University. John Cayer, head of Fairfield University's interlibrary loan system, tracked down numerous articles from engineering journals and books from as early as the 1830s. He always had a smile on his face despite my request for "just one more article."

Of course, I would accomplish nothing without the support of my daughters, Kerry and Kara, a new generation of engaged and curious scholars, and my wife, Dr. Mary Murphy Schlichting, whose keen insights have been crucial in my preparation of the manuscript.

All of us need good mentors in our lives, and I have been fortunate to have encountered two extraordinary people: Arthur Anderson and Richard Maisel. Professor Arthur Anderson, whom I first met as an undergraduate at Fairfield University, encouraged me to think about an academic career. Returning to Fairfield University at the start my academic career, I am proud to have worked for years with Arthur Anderson as a colleague and friend. He will always remain an inspiration to me. Professor Richard Maisel of New York University guided my graduate studies and pointed me to the future. We have remained friends since those graduate school days and continue to do research and publish together.

GRAND CENTRAL'S ENGINEER

INTRODUCTION

ON SUNDAY, FEBRUARY 2, 1913, the *New York Times* devoted an entire special section of the newspaper to the opening of the new Grand Central Terminal on 42nd Street. A lead on the second page hailed the New York Central Railroad for "Solving Greatest Terminal Problem of the Age."[1] The paper celebrated the opening of the New York Central's magnificent new terminal building in the heart of the greatest city in the world. No superlative seemed adequate to extol the grandeur of the new gateway to New York. The *Times* lavished praise on the building's architecture and the soaring space at its core, the Grand Concourse.

The *Times* editorial page, usually given to somber opinions, dramatically referred to Grand Central as "A Glory of the Metropolis" and added:

> Railroad stations possess for a city something of the importance that is possessed for a country by railroads themselves. It is by no means an idle or empty boast, therefore, for New York to proclaim that from to-day it will have in use for itself and its daily army of visitors what are beyond question two railroad stations in every way superior to any other buildings for their purpose in the world. . . . The corporations have expended their millions in no mean and narrow spirit of utilitarianism, but with appreciation of a civic duty to produce architectural monuments to illustrate and educate the aesthetic taste of a great nation.[2]

The second station referred to was the new Pennsylvania Station at 34th Street, opened with similar fanfare by the Pennsylvania Railroad a year earlier. New York City now boasted the two largest and most elaborate rail passenger facilities in the world. What more majestic symbols than Grand Central and Pennsylvania Station could a city that increasingly viewed itself as the most important place on the face of the earth have?

Nowhere in the *Times* coverage can one find the name of William J. Wilgus, the brilliant former chief engineer of the railroad who first presented his revolutionary plans for a completely new Grand Central and an underground electrified train yard to the president and board of the railroad in 1902.

Wilgus was not only ignored in 1913; he remains lost to history. Grand Central, which he undertook at the age of thirty-nine, became the most complicated construction project up to that time in New York City's history. Wilgus proposed taking the railroad's aging Grand Central depot and train yard—stretching from 42nd to 56th Streets and from Lexington to Madison Avenues—and placing the entire rail complex underground, and then building a new Grand Central. Wilgus imagined not just the world-famous terminal building on 42nd Street but also planned both vertically and horizontally, excavating to 90 feet below street level to accommodate a two-story underground train yard. In 1902 no railroad passenger facility in the world incorporated such extensive underground tracks and platforms. Vertically, his plan included hotels, office buildings, and apartments that would use the "air rights" over the underground train yard. All would be linked together by indoor corridors incorporating restaurants and shopping— a "city within a city."

The interwoven arrangement depended upon the successful use of a new technology for motive power—electricity. Wilgus premised the entire project on the electrification of the underground train yard and the design of a completely new class of powerful electric engines. With supreme self-confidence and poise, the new chief engineer convinced senior management and the board of directors, among them the powerful J. P. Morgan, to proceed with the complicated and costly project.

With the new terminal a major section of midtown Manhattan became the most complex passenger transportation hub in the world, with interlocking hotels and office buildings—"Terminal City"—all a part of the vision Wilgus first presented to the railroad in 1902. The new Grand Central complex served as a catalyst to transform the surrounding area into a vibrant midtown business and residential center.

Throughout a long and productive career William Wilgus imagined integrated solutions to New York's complex transportation challenges. He realized intuitively that piecemeal plans and projects brought no real solutions to providing efficient passenger and freight transportation in Manhattan, the rest of New York City, the harbor, and the surrounding metropolitan region.

WILLIAM JOHN WILGUS traced his lineage back to Sussex, England, to the parish of Salehurst circa 1665. His grandparents Alfred Wilgus and the former Lavinia Wheelock, with their five children, moved from Albany westward to Buffalo, New York, in 1827. Their journey by packet boat on the Erie Canal placed them at the heart of the expanding settlement of upstate New York, Ohio, and the Midwest stimulated by the completion of the Erie Canal in 1825. At that time the largest and most complex transportation project in American history, the Erie Canal not only facilitated the westward expansion of the country but also led to the dramatic growth of New York City and its emergence as the leading seaport in the country.

Alfred Wilgus found work as a bookbinder and later opened a bookstore in downtown Buffalo, the western terminus of the canal and for a time among the fastest growing and most prosperous cities in the country. Wilgus's grandparents had twelve children, including his father, Frank Augustus. Four years after

Frank's marriage to Margaret Woodcock, born in Brooklyn in 1841, William John Wilgus was born in 1865 in his parents' home in Buffalo, the same house where his father had been born in 1837.

In his unpublished autobiography Wilgus concludes a chapter on his family's genealogy with a backhanded compliment to his forebearers: "None of those from whom I am descended held places high on the scrolls of fame. . . . Several of them, however, in local history made legible footsteps on the sands of time." He added one of the few truly personal references to himself in the autobiography: "My unfavorable traits, so far as I know, are not to be charged to them, unless it is my touchiness and quick temper and perhaps my tendency to draw into myself."[3] Wilgus's professional life as an engineer, a successful business executive, and a transportation visionary can be reconstructed from the meticulous records he kept over a lifetime of achievement. As a person, husband, colleague, and father he remains elusive.

Wilgus described his father as a modest man of slight build with a nervous temperament whose formal education ended at the age of fourteen, certainly not uncommon in the era before the Civil War, when childhood ended in the early teens and an adult life filled with hard work began. Frank Wilgus went to work for the newly formed New York Central Railroad. Buffalo served as the western terminus for the Central Railroad and later became the heart of the New York Central & Hudson River Railroad empire. The elder Wilgus worked as a modestly paid freight agent at the Central's station in Buffalo for fifty years yet lived to see his son rise to the pinnacle of success with the same company.

The Wilgus family relocated to a 15-acre farm in West Seneca, New York, south of Buffalo, in 1876, when Wilgus turned eleven. Just after the family moved to the farm, a momentous event in Wilgus's life set his course for the future. His father hired Marsden Davey, English by birth and an accomplished civil engineer, to survey the family's newly acquired farm. Wilgus followed Davey around as he surveyed the land "holding the end of his surveyor's chain, peeking through the telescope of his theodolite, and watching him closely as he made a map." This experience proved transformative. Wilgus was fascinated with engineering's "mathematical exactitude . . . and aesthetic charm"—the exact attributes that characterized all of his own work from Grand Central onward. From that moment his future seemed destined: "The year 1876 was, in a way, to me as momentous as 1776 had been to my forefathers a hundred years before. My mind thus was spurred to adopt a settled course from which it never wavered." His career as an engineer commenced.[4]

When Wilgus reached the age of fourteen, his parents enrolled him in Buffalo Central High School, the only high school in Buffalo at that time, even though it was a city of over 150,000 residents. Graduating in the top of his class, eighth out of forty-four students, with an "honor star," he seemed destined to go on to college. Yet Wilgus did not attend a university for the simple reason that his parents could not afford to pay for it.

Even if his father had had the means to provide a college education, few institutions in the country in 1883, the year Wilgus completed high school, offered an engineering degree. The teaching of both engineering and science were limited in the United States. In New York State only four institutions offered training in engineering: the U.S. Military Academy at West Point, Columbia University in New York City, Rensselaer in Troy, and Cornell University's Sibley School of Mechanical Engineering and Mechanical Arts.

Instead, Wilgus's father arranged for him to be apprenticed to Marsden Davey, the civil engineer whom Wilgus had followed around years before as Davey surveyed the family farm. The terms of the apprenticeship, an entirely practical education, lasted two years. The first year Wilgus gave his services to Davey for free; during the second he worked for the modest salary of a dollar a day. He lived at home during his apprenticeship, supported by his father.

Wilgus assisted Davey in surveying undeveloped areas in Buffalo and the surrounding communities for new housing and laying out plans for the local street railway. He developed his drawing skills as he traced building plans in triplicate in an era before blueprinting. In his spare time he continued to study books on engineering and mathematics and completed a correspondence course in mechanical drafting from Cornell University. Wilgus apprenticed with Davey from 1883 to 1885; he left Davey's employ with a letter of recommendation describing him as "qualified to enter the profession in any branch."[5]

At the age of twenty, with Davey's reference in hand, Wilgus decided to move to the West to embark upon his chosen profession. He resolved to succeed in his career despite the fact that he lacked an engineering degree: he determined to overcome this "handicap and win recognition by devotion to duty and good hard work."

Wilgus headed west in the summer of 1885 with sixty dollars in his pocket, taking a twenty-four-hour train ride from Buffalo to Chicago and on to Minneapolis. He carried with him a letter of introduction from his father to a distant cousin, James H. Wilgus. His first hard lesson came with a rejection from the engineering department of the Northern Pacific Railroad. Not to be deterred and with the help of James Wilgus, he secured an interview with the chief engineer of the Minnesota & Northwestern Railroad, Henning Fernstrom, who offered Wilgus a job at fifty dollars a month locating land records to facilitate the construction of the Minnesota & Northwestern south from Minneapolis to Iowa.

Officially employed with the railroad as a rodman and draftsman, Wilgus began a career with the rapidly expanding American railroads that lasted for the next twenty-two years. While in Minneapolis he made the acquaintance of two individuals fated to play a major role in his professional and personal life: a young architect, Charles A. Reed, who became his brother-in-law and one of the architects of Grand Central; and Bion Arnold, destined to become a prominent electrical engineer, whom Wilgus later recruited as a consultant to serve on the Electric Traction Commission charged with planning the Grand Central electrification.

Early on, Wilgus learned a lesson in fortitude. Working with a senior engineer laying out the approaches to the railroad's drawbridge over the Mississippi River, he realized that the engineer's calculations were in error. He chose not to speak up; later, when the error became obvious, he received a severe reprimand. From that moment forward, he decided, "I would frankly express my views when they should happen to differ from those of my superior in rank."[6] Throughout a long career, William J. Wilgus followed this dictate and often found himself at odds with those in power.

Within a year of his employment Wilgus became the designing engineer in charge of the Minnesota & Northwestern terminals in Minneapolis and in charge of the railroad's drafting office. Under the guidance of the chief engineer his practical education continued. He learned to design track layouts, tunnels, and rail yard structures; to supervise construction of a complete railroad system; to prepare contract specifications; and to estimate costs. All of these skills served him well when his career brought him to New York City.

In 1888 Wilgus moved from Minneapolis to St. Joseph, Missouri, where recently installed electrically powered railcars, or "trolleys," ran on the city streets, on one of the first electric street railroads in the country. From St. Joseph he moved in 1890 to Duluth to become locating engineer with the Duluth & Winnipeg Railroad and then returned to Minneapolis and to his earlier railroad employer, now named the Chicago, St. Paul, & Kansas City Railroad, with a salary of $150 a month. His work culminated when the railroad completed its terminal in Kansas City in 1891.

For a few months in 1891 Wilgus worked in Chicago as chief draftsman for the Thomson-Houston Electric Company, soon to merge and become part of the General Electric Corporation. At Thomson, Wilgus worked on the application of electricity as a source of motive power for suburban trolley lines, experience that proved invaluable a decade later when he was planning the electrification for the underground train yard at Grand Central.

Other challenging railroad jobs followed, including serving as locating engineer for the Duluth & Iron Range Railroad, built to open the great iron ore beds in the Mesabi Range in northern Minnesota. Wilgus bushwhacked through the virgin forests of Minnesota with a work party of seventeen to locate the railroad's right-of-way. He returned from the woods with a beard which he grew to offset the impression of being too young to serve as a chief engineer. In his autobiography he adds to his account of the completion of work for the Duluth & Iron Range Railroad a brief reference to his marriage to May Reed, sister of Charles Reed, in March 1892 in Hudson, Wisconsin. Nowhere else in his entire autobiography does Wilgus include another reference to his first wife. His personal life remains in the shadows.

Early success in railroading had taken Wilgus to the West, where the railroads were expanding at a breakneck pace and offered rich opportunities for young, smart, confident engineers. He advanced quickly and completed his practical education in a series of increasingly responsible positions. In each he succeeded brilliantly. In the spring of 1893 Wilgus received an offer of a position as assistant engineer of the Rome, Watertown, & Ogdensburg Railroad in upstate New York, another turning point in his life. He traveled via Chicago to Buffalo to visit his parents and brothers and then on to Syracuse and a connection to Watertown, New York.

The New York Central Railroad leased the Rome, Watertown, & Ogdensburg as part of its giant rail empire stretching from New York City, through New York State, to Chicago and St. Louis. Wilgus's success in rehabilitating the Rome, Watertown brought him to the attention of the Central's senior management in New York City.

Just four years later, in June 1897, Wilgus received an offer to join the New York Central and move to New York City to serve as the engineer for the railroad's eastern division. Six months later another promotion elevated him to the position of chief assistant engineer and then, in 1898, to chief of maintenance of right-of-way. On April 15, 1899, Wilgus, at the age of thirty-four, received still another promotion to the position of chief engineer of the entire New York Central Railroad.

The era from the end of the Civil War to the turn of the century offered abundant opportunity, and no industry offered as much opportunity as the railroads. The railroads transformed American society. From fewer than 31,000 miles of track in 1860 to almost 130,000 miles by 1890, the railroads grew at a staggering pace.[7] While national attention focused on the completion of the first transcontinental railroad in 1869, celebrated with great fanfare in Promontory Summit, Utah, railroads expanded throughout the country. Without the railroads, the industrialization of the United States could not have proceeded as rapidly as it did. The railroads overcame the two challenges provided by the immense geography of the country—time and distance. With the growth of the railroads, transportation costs declined dramatically; and industry in locations close to rail lines could specialize, innovate, and drive down costs by serving an increasingly integrated national market linked by the country's expanding rail system.

Wilgus caught this "railroad fever" when he first went to work for the Minnesota & Northwestern Railroad in Minneapolis in 1885. His salary paid him a modest fifty dollars a month. He seized the opportunities provided to become a railroad engineer from practical experience. Hard work and native ability led to rapid promotion and just fourteen years later, in 1899, he rose to the lofty position of chief engineer and fifth vice president of the New York Central Railroad & Hudson River Railroad, the second largest railroad in the world.

Table I.1. Manhattan population, 1860–1910

Year	Population	% increase from previous decade
1860	813,669	
1870	942,292	+ 15.8
1880	1,164,673	+ 23.6
1890	1,441,216	+ 23.7
1900	1,850,093	+ 28.4
1910	2,331,542	+ 26.0

SOURCE: U.S. Census Bureau, *Census of Population, 1960*, vol. 1 (Washington, DC: Government Printing Office, 1963), part A, table 28.

WHEN WILGUS ARRIVED in New York, Manhattan had about 1.8 million residents, with most of the population concentrated in the lower third of Manhattan Island. In the lower east side wards east of Bowery Street and below Rivington Street, population density exceeded 400,000 residents per square mile, among the most densely populated places on the face of the earth.[8] Between 1880 and 1910, Manhattan's population grew by more than 20 percent each decade (table I.1).

From 1900 to 1910, while Wilgus's new Grand Central Terminal rose on 42nd Street, Manhattan's population increased by 481,449 residents—almost half a million people in one decade. No other city in the United States even approached a similar rate of growth. All of these people were crowded on an island twelve miles long and two miles wide surrounded on all sides by rivers and a harbor to the south. As the population of the city exploded, the demand increased for an efficient transportation system to move people and freight.

New York City now served as the center of the United States' foreign commerce. By 1900 the port of New York had become the busiest port in the world and remained so for the decades to follow. Thousands of ocean-going ships entered the port each year, joined by thousands of coastal vessels serving ports from New Orleans to Portland, Maine. In addition to the cargo ships jamming the harbor and lining the wharfs, millions of passengers arrived by ship each year. The vast majority of passengers were immigrants to the United States destined to settle in Manhattan.

New York also became a center of manufacturing, accounting for over 10 percent of all of the country's industrial output. The 1900 Census of Manufacturing counted 39,776 manufacturing establishments in both Manhattan and Brooklyn, most of them in lower Manhattan or across the East River on the Brooklyn waterfront.[9] Each day tons of freight moved from the railroads and piers to the factories, and then finished goods moved back to the piers or rail yards to be

Bird's-eye view of the tip of Manhattan Island, the New Jersey waterfront, and the crowded New York Harbor, 1892.

distributed locally, sent across the country, or shipped abroad. Horse-drawn drays and, later, commercial trucks carried freight, clogging the city streets as trucks still do today.

MANHATTAN'S UNIQUE GEOGRAPHY creates transportation challenges like no other place in the United States. Walking down the busy, crowded streets in midtown it is easy to forget that Manhattan is, in fact, an island separated from New Jersey by the Hudson River, from Brooklyn and Queens by the East River, and from the Bronx to the north by the Harlem River. To the south of the island stretches Upper New York Bay, separating Manhattan from Staten Island by over two miles. From the city's founding by the Dutch in 1609 to the present day, the water boundaries of Manhattan have created a distinct challenge: providing transportation to and from the mainland. The challenge was how to move people and goods efficiently to and from Manhattan as the city and port of New York grew to be the center of the country's population, commerce, industry, shipping, finance, and culture—the most important city in the country or, as a recent title proclaimed, "the island at the center of the world."[10]

Wilgus devoted his career to solving the perplexing transportation problems resulting from Manhattan's geography. Grand Central constituted only one hub in an incredibly complicated passenger transportation system in New York City and the region.

The new Grand Central Terminal dramatically improved the movement of passengers to midtown Manhattan, but myriad problems remained. Ten of the railroads serving the New York metropolitan region terminated on the New Jersey side of the harbor. Passengers and freight destined for Manhattan and Brooklyn had to be carried over the water by ferry and barge, a complicated, inefficient, and costly transportation system. The shoreline of Manhattan and the Brooklyn waterfront teemed with shipping linking New York with the rest of the world. The piers that lined the Hudson and East Rivers were crowded with ships, ferries, passengers, and freight, creating congestion on the city streets along the waterfront.

New York desperately needed a more efficient system to move both people and freight. The supremacy of the city as the leading port in the country and a manufacturing center hung in the balance, challenges that remain to the present day.

William Wilgus devoted his thirty-year career after the Grand Central project to solving the transportation challenges of the ever-growing city and port. A brief list of his major projects provides a hint at the range of his efforts to improve transportation in New York City, the port of New York, and the region:

- the new Grand Central Terminal complex
- plans for an underground, electrified freight railroad to link lower Manhattan to the railroads across the Hudson River in New Jersey
- plans for an elevated, inter-terminal belt-line railroad around lower Manhattan

- the original idea for a New York–New Jersey "Port Authority" to solve the region's freight transportation problems
- the Holland Tunnel
- plans for a Staten Island–Brooklyn "Narrows" Tunnel and connecting railroad
- plans for a lower deck on the George Washington Bridge for railroad trains
- director of the WPA Work Relief program in New York City during the Great Depression

Time and again Wilgus advocated for an integrated "railroad solution" to New York's transportation problems. He came of age during a period of American history in which the railroads ruled supreme and provided mass transportation systems capable of spanning the country or linking together an ever-expanding metropolitan region surrounding the port of New York with Manhattan Island at its center.

Wilgus struggled with the thirteen competing railroads serving the New York metropolitan region, urging them to cooperate to improve delivery of passengers and freight, and to lower costs. Time and time again he encountered fierce resistance on the part of the railroads to any of his plans that they perceived not be in their vested short-term interest. Politicians in New York City, the port region, and the states of New York and New Jersey also proved to be shortsighted.

As powerful and uncooperative as the railroads remained in the first decade of the twentieth century, another transportation revolution was already under way. In 1900, just three years after Wilgus arrived to work in New York City for the New York Central Railroad, automobile registration in the United States totaled just over 8,000 vehicles and production of automobiles totaled 2,474 vehicles, including the eleven models built by the Oldsmobile Company. Henry Ford formed the Ford Automobile Company in June 1903 and produced the first Model T in 1908. In the following years the floodgates opened; and in 1913, the same year Grand Central opened its doors, Ford produced over 200,000 Model Ts. Soon carmakers brought the motorized truck to the market, and in a very short period of time, both the passenger car and the truck challenged the railroad's dominance in the country's transportation system.

The automobile and truck age would have a profound impact on transportation within the New York metropolitan region. When Wilgus first proposed his revolutionary plans for the new Grand Central complex in 1902, improved rail service provided the only solution to the New York Central's passenger problems in New York City. He projected the city's population would eventually total twenty million residents, with the population density in the outer boroughs matching Manhattan's. Mass transportation was the only practical way to move people and the railroads the only practical way to move freight. Initially Wilgus did not anticipate that the completely new automobile technology would challenge mass transportation and the railroads. Nor could he envision the decentralized spread of the population beyond the boundaries of the city of New York sprawling into the suburbs.

Over the decades which followed the opening of Grand Central, the railroads would have to compete with the auto and the highways for both passengers and freight. Wilgus eventually saw this new world coming and believed the automobile and truck could become an integral part of New York's transportation system. He became the chief consulting engineer for the Holland Tunnel and oversaw the construction of the longest vehicular tunnel in the world, creating the first highway link between Manhattan and New Jersey. Despite the success of the Holland tunnel and the tunnels to follow, Wilgus, to the end of his career, believed the railroad provided a much more efficient means to move both passengers and freight. An integrated system combining efficient rail transportation for the long haul and then auto and truck for the final few miles remained his dream.

By the time he retired in the 1930s it appeared the auto age and the highway age would drive the railroads into oblivion, and many of his plans for improved rail transportation never came to fruition. However, by the turn of the twenty-first century, faith in more highways as the only solution to transportation congestion in New York and the metropolitan region has disappeared. Improved rail and mass transit offer the only solution, just as Wilgus advocated almost one hundred years ago.

— 1 —

NEW YORK CITY'S GEOGRAPHY AND TRANSPORTATION CHALLENGES

WHEN WILLIAM WILGUS arrived in New York City in 1897 to work in the corporate headquarters of the New York Central Railroad, the transportation system of the city and the harbor had grown into a complex, intertwined, and unruly monster. A thriving city and metropolitan region of over twelve million people surrounded the port of New York. Twelve railroads with teeming rail yards crowded the New Jersey shore of the harbor, and the New York Central's two large rail yards on the West Side of Manhattan filled multiple city blocks connected by rail lines on the city streets. Numerous bridges crossed the East and Harlem rivers; and ferry lines from New Jersey and between Manhattan, Brooklyn, and Staten Island crisscrossed the waters. Myriad private companies hauled freight on the crowded city streets in Manhattan and along the Brooklyn waterfront. For the next thirty years Wilgus worked tirelessly in an attempt to bring order and efficiency to the transportation challenges created by New York's geography and the city and region's rise to the pinnacle of American commerce and industry.

New York's Geography

One fundamental geographical fact shapes the fate of New York City and the entire metropolitan region: at the very center of the region sits Manhattan Island, surrounded on all sides by water. The island consists of a narrow slice of land running twelve miles long to the north and less than two miles wide at 14th Street and much narrower as one moves north. To the west, the Hudson River, or North River, forms a mile-wide boundary separating Manhattan from New Jersey. To the east, the East River separates Manhattan from Brooklyn and Queens, both part of Long Island, which extends a hundred miles to the east to Montauk Point. To the north, the Harlem River separates the island from the Bronx and Westchester.

The siting of an urban area on an island provides a complicated challenge. Cut off from the mainland, an island's population and economy always remain dependent upon connections to the larger world, whether the island is a small place with few residents (like Block Island, approximately 120 miles to the east of New York City) or a densely populated place like Manhattan. Historically, the overwhelming geographical challenge for Manhattan Island and the city of New York

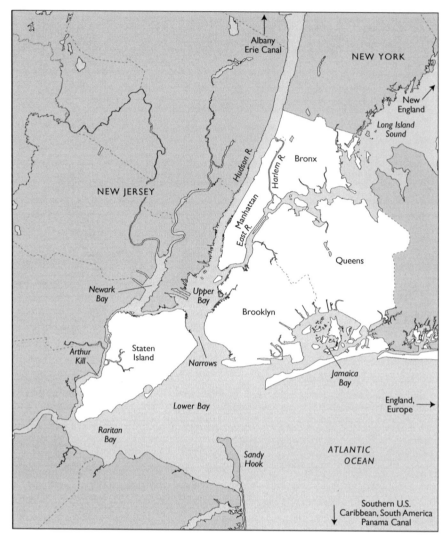

New York Harbor and Manhattan, "the island at the center of the world." The Hudson River leads to Albany, the Erie Canal, and the Midwest; the East River to Long Island Sound and New England; the Atlantic Ocean to Europe, the southern states, South America, the Panama Canal, California, and Asia.

has been to create a system of communication with the surrounding mainland to adequately serve one of the most densely populated places in the world—for well over 350 years a challenge never adequately met.

The early Dutch settlers of New York built an economy centered on the fur trade with the Native Americans. The needs of early residents were few, and communication by boat proved adequate. Manhattan's first European settlement, in the 1600s, located at the very southern tip of the island, had a small population and modest needs. Food could be obtained from the few farms on the island. Communication with Amsterdam, across the Atlantic Ocean, proved to be far more important than easy travel to Brooklyn or to the New Jersey side of the harbor. But as soon as the city's population grew, travel on and off the island dramatically increased in importance.

The second geographical feature that has been critical to New York's history is the superb harbor just to the south of Manhattan Island. Surrounded on all sides by land, the Upper Bay forms a large estuary sheltered from the Atlantic Ocean to the south. For over four hundred years, the Upper Bay has provided ships a perfect shelter from ocean storms. When Henry Hudson sailed into the Upper Bay in 1609, he immediately recognized what an important harbor the geography of the bay and the surrounding land created. The harbor served as the cornerstone for the spectacular growth of both New York City and the region. But the Upper Bay forms only one part of a complex harbor connected, southward, to Lower New York Bay and the Atlantic Ocean beyond.

The Upper Bay joins the Lower New York Bay by the Narrows, the body of water between Brooklyn and Staten Island. Beyond the Lower Bay lies the Atlantic Ocean, providing access to the world's commerce. Staten Island, on the west side of the Narrows, is separated from New Jersey by the Arthur Kill and the Kill van Kull, narrow waterways with names dating back in time to the early Dutch settlers. Two major rivers join at the tip of Manhattan Island, the Hudson and the East River. The East River forms a tidal estuary running north, parallel to the Hudson River to join Long Island Sound through the Hell Gate. The passage up the East River, through the Hell Gate, and to Long Island Sound forms an "inland passage" for shipping to New England. With the opening of the Cape Cod Canal in 1916, ships could travel from New York City to Boston sheltered most of the way from the Atlantic Ocean. The Hudson River extends northward for over 300 miles and is navigable for 130 miles to Troy, New York, providing direct water access to upstate New York.

The New Jersey shore, opposite Manhattan, forms the Bayonne peninsula. Further west of the peninsula sits Newark Bay; the Hackensack and Passaic Rivers empty into the bay. Newark Bay and the city of Newark today serve as the center of the port's shipping, with vast container piers and convenient rail and truck access. The region's shipping activity, once centered on Manhattan Island and the Brooklyn waterfront, shifted to Newark Bay in the 1960s and '70s, and the once bustling Manhattan and Brooklyn waterfronts faded. The great piers lining the Hudson and East Rivers fell into decay; now only the occasional cruise ship or ocean liner berths, with great fanfare, at the once busy piers along the shore of Manhattan Island.

Lower New York Bay, to the south of Staten Island, bounded by Sandy Hook on the east, forms a sheltered anchorage from the Atlantic Ocean and the prevailing wind from the southwest. To the west, Raritan Bay and the Raritan River, navigable to New Brunswick, New Jersey, once served as an important water route between Philadelphia and New York. The Delaware and Raritan Canal, completed in 1834, linked New Brunswick with the Delaware River at Bordentown, New Jersey, and then to Philadelphia further south down the Delaware River.

With over eight hundred miles of shoreline, the port of New York ranks among the greatest natural harbors in the world. Along the shoreline of Manhattan Island and Brooklyn, relatively deep water provided access to piers lining the shores. In an earlier era, when ship-borne commerce provided the lifeblood of a mercantile economy, seaports along the Atlantic played a crucial role in the U.S.

economy, no port as dominant as New York. Robert Albion, a historian of the port, identifies the harbor's key natural advantages: "New York possessed a land-locked harbor, which offered more perfect natural shelter than that of any other major American port as close to the sea. A century ago, a prominent engineer remarked that even if the sandy mass of Long Island might have no other use, it justified itself as a natural breakwater for New York. Staten Island served similar purposes. . . . Shipping in the Lower Bay or the Narrows might catch the force of a gale, but the waters around Manhattan were spared the surges of stormy seas."[1]

Around the Upper Bay, the city of New York grew and gave birth to an entire metropolitan region. Albion defined the port region as a twenty-mile radius around the tip of Manhattan Island.[2] William Wilgus used a similar geographical framework, as did the first planners for the Port Authority of New York and New Jersey. A twenty-mile radius from Manhattan encompasses the Bronx and the lower part of Westchester County to the north; Brooklyn, Queens and eastern Nassau County to the east; Staten Island and the Atlantic Ocean to the south; and, on the New Jersey side of the harbor, Bergen, Essex, Hudson, Middlesex, Passaic, and Union Counties. Extending the radius to fifty miles defines the modern tri-state metropolitan region, which has a population of almost twenty-one million people today.

The History of New York

The first Europeans arriving in North America settled along the coast, with immediate access to the sea. The Jamestown adventurers chose a location up the James River from Chesapeake Bay that offered security from the perceived threat from the Spanish to the south but fronted directly on the river, providing a safe anchorage for their ships. At Plymouth, Massachusetts, the Pilgrims located their settlement on the shore of a small bay with access to Massachusetts Bay and the Atlantic Ocean beyond. All of the early settlements in North America remained tied to the sea routes linking the new colonies to Europe.

In March of 1524 the explorer Giovanni de Verrazano, sailing under the flag of Francis I of France and searching for a sea route to the riches of China and Japan, anchored in the Upper Bay just south of Manhattan Island. Rumors of other European explorers' finding their way into New York harbor followed over the succeeding years until the *Half Moon*, under the command of the Englishman Henry Hudson, also searching for a passage to the Orient, sailed into the harbor in September of 1609—eighty-five years after Verrazano. In the employ of the Dutch East India Company, Hudson explored the river opening to the north on the west side of Manhattan Island seeking the magical water passage to Asia. Over one hundred miles to the north, near present-day Albany, the river shallows prevented further passage, and Hudson's search for the fabled "northwest passage" through the North American continent to Asia ended in disappointment.

The Dutch East India Company, despite Hudson's failure to find the rumored passage to the Indies, realized that New York harbor and the Hudson River offered an opportunity for commercial gain—commerce that would not involve shipping the riches of China to Europe but, instead, trading furs abundant in the forests of this new world. The company, based in Amsterdam, sent additional ships in the years following Hudson's voyage to the protected harbor of New York. In 1624, the company established the settlement of New Amsterdam at the tip of Manhattan Island and also sent a small number of settlers further up the Hudson River at Albany. From this small group of Dutch pioneers, the city of New York arose.[3] In 1626 a Dutch ship bound for Amsterdam cleared New York harbor with a cargo of 7,244 beaver skins, 1,000 other animal skins, and oak and hickory barrel staves. New York's commercial links with the world began with this modest shipment of furs and wood to Holland.

A long, protracted struggle for dominance in North America between the Dutch and English led to the surrender of New Amsterdam to the English on September 8, 1664. Governor Peter Stuyvesant marched out of Fort Amsterdam at the tip of Manhattan Island, and the English assumed control, with Richard Nicolls serving as first English governor. Henceforth the colony would be called New York in honor of the Duke of York, the younger brother of Charles II. The king granted York and his supporters all of the territory from the Connecticut River to the Delaware River in exchange for recognizing Charles as sovereign.[4]

Under English rule New York prospered. In the colonial period before the Revolutionary War, Manhattan remained an important port among the English colonies along with its main rivals: Boston, Baltimore, Charleston, Newport, and Philadelphia. Trade between England and the colonies increased steadily, with Virginia and Maryland dominating with the export of tobacco. New York trailed in exports but increasingly served as a port of entry for imports from Britain (table 1.1).

The Revolutionary War devastated New York. After defeating the American forces in a series of battles fought in Brooklyn and Manhattan in 1776, the British occupied New York and did not leave until 1783, after seven long years of occupation. During the British occupation, the city's economy ground to a standstill. Estimates recorded a population decline from over twenty thousand residents in 1776 to fewer than ten thousand when General Washington triumphantly entered the city on November 25, 1783, as the last departing British soldiers boarded transport ships lying in the East River.

Rise to Prominence

New York recovered rapidly after the Revolutionary War. After the Revolution, Britain remained the new country's main trading partner despite the recent war, and New York served as the key port

Table 1.1. Trade between Britain and its American colonies (in pounds sterling)

Year	Virginia and Maryland	New England	New York
	EXPORTS		
1697	227,756	26,282	10,093
1727	421,588	75,052	31,617
1757	418,881	27,556	19,168
1774	612,030	112,248	80,008
	IMPORTS		
1697	58,796	68,468	4,579
1727	192,965	187,277	67,452
1757	426,687	363,401	353,311
1774	528,738	562,476	437,937

SOURCE: Robert Albion, *The Rise of the Port of New York* (New York: Charles Scribner's Sons, 1939), 117.

Table 1.2. Exports and imports through the largest U.S. ports, 1821–1824 (millions of dollars)

Year	Total U.S.	New York	Boston	Philadelphia	Baltimore
			EXPORTS		
1821	64	13	12	7	3
1822	72	17	12	9	4
1823	74	19	13	9	6
1824	75	22	10	11	4
			IMPORTS		
1821	62	23	14	8	4
1822	83	35	18	11	4
1823	77	29	17	13	4
1824	80	36	15	11	4

SOURCE: Albion, *Rise of the Port of New York*, 390–91, app. 2.

linking the United States and Britain. New York's commercial relationship with Britain's rising textile industry drove the city's commercial and shipping growth. The industrial revolution in Britain, centered on the manufacture of textiles, created an insatiable demand for cotton and to a lesser extent wool. Soon the United States was Britain's main source of cotton. For over a hundred years, until the Civil War, the export of cotton to Britain and the import of textiles in return dominated American overseas trade.

New York's overall share of the nation's overseas commerce continued to increase.[5] Albion documented New York's increasing dominance over its chief rivals—Baltimore, Boston, and Philadelphia—even before the opening of the Erie Canal. By 1824, 29 percent of the country's exports and 45 percent of imports passed through the port of New York (table 1.2).

In addition to trade, New York also emerged as a center of the ship-building industry. Up the East River at Corlear's Hook, from Montgomery Street to 14th Street, thirty or more ship yards built many of the ships famous for spreading American commerce around the world. The famous Brown & Bell's yard, at the end of Houston Street, at one point employed over five hundred shipwrights, caulkers, blacksmiths, and riggers. Farther up the East River, between 5th and 7th Streets, the firm of Webb and Allen, later William H. Webb, constructed some of the most famous ships ever to be built in New York. An article in *Harper's New Monthly Magazine* described the view from the river itself: "The traveler who sailed down the East River and saw the yards that lined the New York shore, the noble vessels on the stocks, the thousands of busy workman, and the huge stock of timber—white oak, hackmatack, and locust for the ribs of the ships, yellow pine for the keelsons and ceiling timbers, white pine for the floors, live-oak for the 'aprons'—might have been pardoned for supposing that Manhattan Island was the head-quarters of the ship-building of the world: for such indeed it was."[6]

Brown & Bell built many famous ships, including the *Liverpool* and the *Queen of the West* (1843), which sailed for the Liverpool packet lines. Packet ships revolutionized transatlantic travel and shipping because they departed for Europe on a fixed schedule. Regardless of the number of passengers or the amount of cargo loaded aboard, a packet ship departed on a specific day of the month and at the appointed hour. Previously, ships had left New York or any other port at the whim of the ship owner or captain, who wanted his ship full of cargo before leaving port. While the packets could not guarantee the time to cross the Atlantic, they dramatically improved communication between the United States and Europe and vastly aided the increase in transatlantic trade.

In October of 1817 advertisements appeared for four ships—the *Amity, Courier, Pacific, and James Monroe,* all built in New York—that would, beginning in January 1818, sail on a specific day each month to Liverpool, England. These fours ships represented the founding of the famous Black Ball Line, the first packet line between New York and Liverpool. The Red Star and Blue Swallowtail lines followed in 1822. The ships of the Black Ball Line had a large black ball imprinted on their foresails aloft for all to see as they raced down the harbor toward the Atlantic and England. The Black Ball Line continued to operate service to England for over sixty years with ships built in the East River yards.[7] By 1825, twenty-five packets operated from the port of New York, and the number increased to forty-eight by 1840 and fifty-two by 1845. With the number of packet ships increasing, passengers and shippers had a choice of two or three ships leaving for Europe each week. No other American port ever approached this level of service, and the packet lines further increased the city's lead as the country's premier port.

New York harbor also witnessed the birth of the steamboat era in the United States. On August 17, 1807, Robert Fulton piloted his steamboat, the *Clermont*, from New York up the Hudson River at the leisurely speed of five miles an hour. Built by the Charles Browne yard on the East River, the *Clermont*'s steam engine came from England. Soon American manufacturers in Manhattan began building steam engines. Steamboats dramatically reduced the travel time on the important route between Manhattan and Albany and created faster ferry service across the rivers and throughout the harbor. The *Clermont* advertised a time of thirty hours for the 150-mile trip to Albany for a fare of seven dollars, a trip that averaged from three to nine days by sail. Steamboat service centered in New York harbor grew substantially, offering regular, scheduled service with a predictable time schedule, an enormous advance over sail, which was subject to uncertain winds and powerful currents on the Hudson and East Rivers and Long Island Sound.

The Erie Canal

On November 4, 1825, Governor DeWitt Clinton of New York and a host of dignitaries aboard a flat-bottomed canal boat, the *Seneca Chief,* poured two casks of water from Lake Erie into the Narrows between Brooklyn and Staten Island. The *Seneca Chief,* having left Buffalo ten days earlier, had just traveled 360 miles on the just-completed Erie Canal to Albany and then down the Hudson River to New York City. A parade and celebration followed in Manhattan with an estimated one hundred thousand spectators—"nearly two-thirds of the city's population . . . the largest gathering ever witnessed in North America."[8]

The Erie Canal, a dream of New Yorkers for more than a hundred years, constituted the largest and most complex construction project in American history up to that point and was the longest canal in the world. The canal took advantage of a crucial geological fact. From northern Alabama to Maine's border with Canada stretch the Appalachian Mountains, creating a natural barrier separating the Atlantic seaboard from the interior of the North American continent. While not as rugged or elevated as the Rocky Mountains, the Appalachians rise to significant heights, preventing easy transportation between the eastern United States and the Midwest. Only in one place does a relatively level path lead from the East to the Midwest—upstate New York. Beginning in Albany, the Mohawk River Valley extends almost ninety miles to the west; and beyond the Mohawk Valley, Lake Oneida and a series of low, rolling hills and valleys complete the path to the shore of Lake Erie. The Erie Canal took full advantage of this "water-level" route across upstate New York.

The opening of the Erie Canal for its full length between Buffalo and Albany in 1825 created a transportation revolution between the Midwest and the eastern seaboard and provided a decided advantage to New York City and its harbor. Shipping costs declined dramatically, and the ever-increasing bounty of the Midwest began to be shipped by the canal to the port of New York. Canal traffic and revenue increased each year until the railroads in New York State began to offer serious competition in the 1850s.

From the beginning the canal's toll revenue exceeded expectations. In the first full season of operations in 1825, "an average of 42 boats a day passed through Utica, carrying a thousand passengers a day. The eastbound cargoes totaled 185,000 tons, including 211,000 barrels of flour, 562,000 bushels of wheat, 435,000 gallons of 'domestic spirits,' mostly whiskey, and 32,000,000 board feet of timber."[9] A flood of freight and passengers ensured the financial success of the Erie from the day the canal opened. Toll revenues increased from just less than $500,000 in 1825 to over $1 million by 1831, a 100 percent increase in just six years.[10]

Eastbound traffic between Buffalo and Albany exceeded westbound, reflecting the growth of agricultural exports from Ohio and the states to the west. Westbound freight consisted primarily of manufactured goods. However, westbound traffic also included significant numbers of passengers moving to a new promised land in upstate New York, Ohio, and beyond to the west. William Wilgus's grandparents and their five children traveled on the Erie Canal to Buffalo in 1827 to find new opportunity in this rapidly growing city, the gateway to the canal from the Midwest and Canada.

Between 1835 and 1850, traffic over the Erie Canal continued to increase

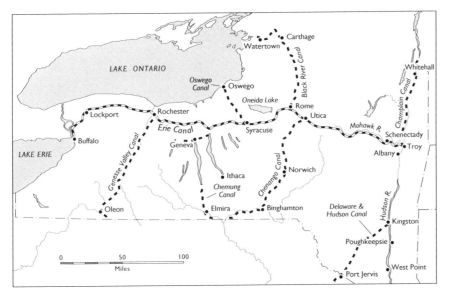

New York State
canals, 1860.

dramatically even with the building of the first railroads across New York State. A virtual cornucopia of agricultural products arrived at Buffalo and was then transported across to Albany and down the Hudson to the port of New York, much of this agricultural bounty destined for export (table 1.3).

William Cronon's masterwork about Chicago, *Nature's Metropolis*, explains the dramatic rise to dominance of Chicago over the Midwest's grain, lumber trade, meat packing, and business in general. In turn, a key to New York City's rise to commercial dominance rested on its links to Chicago: "American boosters saw London as current seat of world empire, heir to Rome's throne, but they also believed that New York would soon win the title for itself. . . . The most important factor creating the next imperial metropolis would be the western trade of North America and that New York's primacy depended on that trade. . . . New York dug the Erie Canal to make itself the metropolis of the Great West."[11] An internal "triangle trade" developed, with farmers shipping their products to Chicago and then from Chicago, over the Erie Canal, to the port of New York. In return, New York manufacturers and importers shipped their merchandise directly to Chicago for distribution throughout the Midwest.

New York's symbiotic relationship with Chicago lasted beyond the canal era. Once the railroads replaced canals as the primary means of transportation, the race to link the Atlantic coast to the Midwest by rail focused on Chicago. All of the major eastern trunk-line railroads—the Baltimore & Ohio, the Erie, the Pennsylvania, and the New York Central—competed vigorously to create rail links to Chicago to secure the transportation to and from Chicago for both freight and passengers. The New York Central capitalized on its water-level route across upstate New York and its direct rail access to Manhattan Island for both passengers and freight.

Above all else, the Erie Canal solidified the rise of the port of New York to a position of commercial dominance. As the business historian Ronald W. Filante

Table 1.3. Erie Canal imports at Buffalo from Ohio, Indiana, Illinois, and the West, 1835–1850

Product	Unit	1835	1840	1845	1850
Flour	barrels	66,223	633,790	171,468	984,430
Wheat	bushels	98,071	881,192	1,354,990	3,304,647
Corn	bushels	14,579	47,885	33,060	2,608,907
Provisions	barrels	6,562	25,070	68,000	146,836
Potash	barrels	4,419	7,068	34,662	17,504
Barrel staves	number	2,585,273	22,410,060	88,296,431	159,479,564
Wool	pounds	140,911	107,794	2,957,761	8,805,817
Butter, cheese	pounds	1,030,633	3,429,687	6,579,097	17,534,961

SOURCE: James L. Barton, *Commerce of the Lakes and Erie Canal: Its Nationality of Character* (Buffalo: Beaver's Power Presses, 1851), 5.

summarizes, at its "peak in 1853, the value of goods shipped by the canal from the West exceeded $135,000,000."[12] The "West" referred to the five states of Illinois, Indiana, Michigan, Ohio, and Wisconsin—a huge hinterland. Chicago dominated this region and consolidated much of its trade through the city, as Cronon illustrates. In turn, Chicago merchants shipping directly to New York by the Erie Canal, and later the railroads, ensured that the port of New York handled the majority of the country's import and export trade. On the piers lining the East and Hudson Rivers in Manhattan and along the Brooklyn waterfront, the bounty of the Midwest arrived for export. In Red Hook, on the Brooklyn waterfront, a wealthy New Yorker shipper, William Beard, constructed the immense Erie Basin dock facility in the 1850s. At the Erie Basin canal barges towed down the Hudson River transferred their wheat and corn.

Ships crowded the harbor throughout the year. Edwin Burrows and Mike Wallace note that "in 1849 over three thousand ships sailed or steamed into the harbor from more than 150 foreign ports—three times the number that had arrived in 1835."[13] Three thousand ships a year—a hundred ships a day—filled the Upper Bay and the piers along the shoreline. The Erie Canal played a major role in the success of the port of New York by providing the crucial transportation link with the booming Midwest in the era before the railroad age.

The Cotton Triangle

To the south of New York and along the Gulf Coast, the "cotton" ports of Charleston, Savannah, Mobile, and especially New Orleans competed with New York for overseas trade. These ports completely depended on one agricultural export: cotton. The export of cotton to Europe—overwhelmingly to Liverpool, England—supplied the cotton textile mills in Britain with the raw material they depended upon. England had no domestic supply of cotton; all of it had to be imported before the mills in Liverpool or Manchester could spin the raw cotton into thread and then weave the thread into cloth. England then exported cotton products to the rest of the world, earning handsome profits.

Never far from any study of antebellum trade and especially the trade in cotton lurks the history of American slavery. At its very core, the cotton trade depended upon the labor of millions of slaves. No dry statistics on export trade to Britain or the euphemism of the "cotton triangle" can avoid the fact of human bondage. From 1824 to 1869 a major share of the entire exports of the United States consisted of raw cotton grown on Southern plantations by slave labor. In 1824 exports from New Orleans totaled $7 million. By 1860, on the eve of the Civil War, these exports were fifteen times as much—$107 million—and accounted for 90 percent of New Orleans's exports.

New York City and its shipping interests benefited directly, as eventually a significant share of the cotton export trade passed through its port. Some cotton exporters found it more convenient and cheaper to ship cotton north to New York on coastal vessels, offload the cotton bales on the wharfs along the Hudson and East River, and then reload on ships bound for Liverpool. In 1860, cotton exports from the port of New York totaled $12.4 million, surpassing exports of Erie Canal–transported wheat and flour, which accounted for $8.9 million.[14] Even in New York, cotton remained king.

Stevedores unloading cotton, East River pier, 1877.

The New England Triangle

New York's involvement in the cotton trade continued with the rise of the American domestic textile industry, centered in New England. In the United States, the industrial revolution began with Samuel Slater's mill in Pawtucket, Rhode Island, opened in 1790. Soon, along the rivers in Rhode Island, in close proximity to the port of Providence and Narragansett Bay, cotton textile mills harnessed water power to spin cotton into yarn and weave cloth.

Throughout New England the cotton textile industry expanded dramatically. Between 1816 and 1840 employment in the textile mills increased from fewer than five thousand workers to over one hundred thousand, accounting for one-seventh of total employment in New England, where agriculture still remained the primary industry. A recent study summarized the importance of textile manufacturing: "By 1849 textile manufacturing accounted for two-thirds of the value added by all large-scale New England industries, and by 1860, cotton manufacturing had become the leading industry in the United States as measured by the amount of capital employed, and the net value of the product produced."[15]

All of the cotton processed by the mills in Rhode Island and New England came from the South, and a significant portion, especially the cotton destined for the mills in Rhode Island, passed through the port of New York. On the Hudson River, the Savannah Line and the Old Dominion Line occupied piers below Canal Street, providing freight service between New York and the southern cotton ports. On the East River the Mallory Line, with service to New Orleans, occupied piers 15 and 16 adjacent to the Fulton Street Ferry. Horse-drawn drays transferred the cotton to the Providence Line and the famous Fall River Line on the East River piers at Murray and Warren Streets just north of the Battery. Loaded onto the steamers bound to Providence and Fall River by way of the East River, Long Island Sound, and Narragansett Bay, the cotton arrived on the waterfront in Providence or Fall River to be delivered to the Steam C Manufacturing Company, Randall's Mill, and the one hundred other textile mills in the Providence and Pawtucket region in Rhode Island. Of course, the cost of unloading, hauling, and reloading the cotton in New York contributed to the city's commercial vibrancy.

New York eventually captured the major share of the transshipment of cotton to the New England textile mills. In the 1836–37 season, coastwise shipping from New Orleans carried 39,000 bales of cotton to Boston, most destined for the textile miles in Lowell, Massachusetts, and the rest for mills in New Hampshire and Vermont. New York received 23,000 bales, with Philadelphia far behind with 6,000. Six years later, New York received the major share: 82,000 bales, followed by Boston with 72,000.[16]

Transatlantic Rivalry

While cotton accounted for a major share of the total exports of the United States until the Civil War, the race to create fast, reliable passenger and freight connections to and from Europe also contributed to the dominance of the port of New York. As early as 1818, packet ships sailing from New York provided the first scheduled service to England and the Continent. During the age of sail, ship yards lining the East River built some of the most famous sailing vessels of all time—the great "clipper" ships. Powered by three or four masts with up to twenty square-rigged sails, clipper ships sailed the seas at speeds of up to 19 knots, and their fame echoes across time.

While the New York newspapers celebrated the exploits of the clippers, British shipbuilders began to equip their vessels with steam power to augment their sails. The first steam-powered ships combined sails with a paddle wheel suspended over one side which would be employed only when the wind diminished. British shipping companies continued to innovate, and in 1850 the first screw propeller ship crossed the Atlantic and iron hulls replaced wood. Between 1850 and the beginning of the Civil War in 1861, the proportion of freight carried across the Atlantic in steamships increased from 14 to 28 percent as the era of sail faded.

Among the first British transatlantic steamship lines, operating paddle-wheel steamers and then screw propeller ships, one name stands apart: Cunard. The company's founder, Samuel Cunard, born in Halifax, Nova Scotia in 1788, began his transatlantic shipping company in 1840, building four wooden ships powered by steam-operated paddle wheels with sails for auxiliary power. One of these, the *Britannia*, made the trip from Liverpool to Boston in the summer of 1840 in fourteen days with a stop in Halifax. Cunard's success depended upon a contract with the British government to deliver mail to the United States on a regular basis—at first, twenty round trips a year—providing the Cunard line with a crucial subsidy as the company prospered over the years.[17] In 1845, the British government increased Cunard's mail subsidy to provide for weekly sailings, with half of the trips to end in New York. Cunard's pier 40 on the Hudson River at Houston Street hummed with activity whenever a ship arrived and its first-class and steerage passengers disembarked onto crowded West Street. The fame of the Cunard line continued to the end of the twentieth century with famous ships such as the *Queen Mary* and *Queen Elizabeth*.

Just up West Street at pier 45, the ships of another British steamship company, the White Star Line, berthed. The White Star Line also sought to attract the cream of transatlantic passengers and high-value freight by building faster and more luxurious ships, culminating with the famous *Titanic*, which tragically sank on its maiden voyage to New York in 1912 with the deaths of over fifteen hundred passengers and crew.

"Commodore" Vanderbilt and the Port of New York

The story of the rise of the port of New York, the advent of the railroad age, the opening of the new Grand Central in 1913, and William Wilgus's career all intertwine. The history of Grand Central's parent company, the New York Central & Hudson River Railroad, can be traced back to 1810 and a small sail-powered boat offering to transport freight and passengers across the Upper Bay from Staten Island to Manhattan. Cornelius Vanderbilt, later known as the Commodore, born on May 27, 1794, began his career in transportation when he borrowed a hundred dollars from his parents and purchased a small sailing vessel called a periauger and started a ferry service from Stapleton on Staten Island to the Battery at the tip of Manhattan Island. Rather than operating a regularly scheduled ferry service, in summer months Vanderbilt, just sixteen years old, hoped for at least one round trip a day hauling passengers and whatever freight could be fit aboard.

Vanderbilt's small business prospered, and he soon ordered a larger boat. In 1813 he commissioned the *Swiftsure*, a sloop, 65 feet in length, from a yard located on the Passaic River. The new boat represented a financial risk; the nineteen-year-old sank all his earnings from his much smaller periauger into the new vessel. A born risk-taker, Vanderbilt often made bold economic moves in a career that led to one of the largest fortunes in American history. Other boats followed as Vanderbilt expanded his ferry business into Long Island Sound, up the Hudson River to Albany, and entered the coastal trade as far south as Charleston and Savannah.[18]

 By the start of the Civil War, the Commodore's fortune from his steamship business, conservatively estimated, totaled $20 million, placing him among the richest individuals in America.[19] Shrewd as ever, Vanderbilt began in the 1840s to turn his business interest to the new railroads, which offered opportunity to those willing to gamble on a fledgling industry. In this, he was "like so many early steamboat men . . . [who] transferred [their] principal interest from steam transportation on the water to that on land."[20]

Vanderbilt's railroad interests centered on the first railroads built to provide service to Manhattan Island. In 1849 he began to speculate in railroad stock and obtained a controlling interest in the New York & Harlem Railroad. The "Harlem," as it came to be known, held a franchise of immense value. In 1831, the company received a charter from the state of New York to construct a rail line on Manhattan Island, using 4th Avenue as its right-of-way, to connect the city (lower Manhattan Island) with the village of Harlem to the north next to the Hell Gate on the East River. The railroad completed a line to Harlem in 1837 and then into the Bronx and on to White Plains in Westchester County in 1844. Eventually, the Harlem extended its line to Chatham, near Albany, a modest distance of 149 miles north of the city.

The Harlem never made much money, but to Vanderbilt that was beside the point. The Harlem possessed an asset of immense value: the right to bring its trains onto Manhattan Island. In addition, along 4th Avenue (soon to be Park Avenue) at 42nd Street, the Harlem purchased land for a train yard and wood lot to service its steam engines. On this land the grandsons of Commodore Vanderbilt built the new Grand Central Terminal and underground train complex envisioned by William Wilgus. A direct line runs between the very modest beginnings of Vanderbilt's career with his small sailboat ferrying passengers between Staten Island and Manhattan in 1810 and the magnificent Grand Central Terminal at 42nd Street, the centerpiece of the Vanderbilt railroad empire, opened just over a hundred years later, in 1913.

As the Civil War approached, the Commodore, already in his sixties, continued to expand his railroad interests. In addition to the Harlem, he gained control of the Hudson River Railroad, which had been chartered by the state of New York in May 1846 to construct a rail line down the east side of the Hudson River linking Poughkeepsie to New York City, 74 miles to the south. Service from Poughkeepsie to New York began on December 31, 1849.

The Hudson River Railroad also had the right to bring its trains onto Manhattan Island and down the west side almost to the southern tip of the island at Bowling Green. In addition, the railroad's charter included the right to lay tracks on the city's streets. From 61st Street south, the railroad constructed tracks on 12th Avenue, 11th Avenue, West Street through Greenwich Village, and then down West Street south to St. John's Park just to the west of City Hall. With tracks running down the west side of Manhattan, the Hudson provided direct freight service to the piers lining the Hudson River. As the port of New York captured the lion's share of the country's imports and exports, the Hudson River's freight business prospered. The railroad also constructed two large train yards, one at 61st Street and the second at 34th Street, with accompanying freight depots adding congestion to the surrounding streets.

Over time, objections to the railroads' use of the city's streets grew, especially to the Hudson's tracks on 11th Avenue, soon to be known as "Death Alley" for the number of pedestrians killed by the Hudson's freight trains. For decades the city of New York fought to compel the railroad to remove its tracks from the city streets. Not only did the railroad refuse to remove tracks as the port of New York prospered, but it also increased the number of freight trains on the city's streets, creating the perennial "West Side Problem."

Cornelius Vanderbilt and his son William Henry continued to expand the family's railroad empire. Just after the end of the Civil War, the Vanderbilts turned their eyes to the New York Central Railroad, whose main line ran from Albany to Buffalo, paralleling the route of the Erie Canal.[21] Through a series of deft financial machinations, the Vanderbilts gained a controlling interest in the New York Central, and in 1867 the Commodore became president of the railroad and William Henry vice president. Two years later, the Vanderbilts "persuaded" the state legislature to approve the merger of the New York Central with their Hudson River Railroad to form the New York Central & Hudson River Railroad, the official corporate name of the railroad with direct service to Manhattan Island. Next, the Central leased the Harlem River Railroad and integrated the Harlem's operations with the Central, consolidating control of all direct rail access onto Manhattan Island.

The Vanderbilts decided to use the Hudson River's tracks on the west side of Manhattan solely for freight traffic. With the dramatic growth of shipping using the piers on the Hudson River, the railroad continued to earn strong profits from hauling freight. The Commodore next decided to have all of his railroads' passenger trains use a new terminal on the east side of Manhattan at the Harlem's

New York Central grain elevators at the 60th Street train yard on the Hudson River and Erie Canal barges being towed down the river, 1877.

rail yard at 42nd Street. He commissioned the architect Henry Snook and the engineer Isaac Buckhout to design "Grand Central Depot," which opened in 1871. Grand Central Depot included a large train shed to the rear of the terminal building facing 42nd Street, the largest enclosed space in the country. As the years passed, the depot became an outmoded, crowded, and inefficient passenger facility; and thirty years later William Wilgus brilliantly planned its replacement, Grand Central Terminal. Along the Hudson River, however, the New York Central's freight operations remained chaotic. After the Grand Central project, Wilgus worked on a series of plans to solve the "west side problem" created by the use of the city's streets and waterfront by the freight trains of the Central Railroad.

Immigrant New York

Not only did the port of New York rise to commercial dominance, but the city's population exploded, contributing to the never-ending need for improved transportation. New York became the "golden door" for millions of people coming to America. From 1820 to 1924, during the "century of immigration," 36 million immigrants arrived in the United States; and two-thirds of them passed through the port of New York.[22] No other American seaport—not Baltimore, Boston, or Philadelphia—ever came close to New York in the scale of immigrants arriving.

In just the five-year period of the Irish famine, 1849–54, a total of 1.7 million immigrants arrived in New York. Some immediately left for other destinations, traveling up the Hudson River to Albany to board Erie Canal boats for the western states; and others traveled by train to distant parts of the United States. But many stayed in the city, settling in the teeming immigrant neighborhoods of Manhattan and Brooklyn. The 1860 census reported 47 percent of Manhattan Island's population to be foreign-born, as was 39 percent of Brooklyn's.[23] By comparison, just over 13 percent of the total U.S. population was foreign-born.

Until the 1850s, immigrants arriving in New York simply left the ships on which they had traveled, tied to piers lining the East and Hudson Rivers or anchored in the harbor, and joined the bustling cityscape. In an attempt to bring order out of chaos, the New York State legislature in 1847 established the Board of Commissioners of Immigration of the State of New York. In 1855, the commissioners secured the use of Castle Garden, originally built as a fort at the Battery to protect New York from attack from the sea. At Castle Garden, at the new Emigrant Landing Depot, officials guided the immigrants through a formal registration process and then worked with licensed boardinghouses and railroad agents to protect the new arrivals from the worst abuses.

During the period from 1855 through 1869, staggering numbers of immigrants arrived each year at Castle Garden (table 1.4). Even during the Civil War years, when immigration declined, hundreds arrived at the Battery each day. After the war ended, immigration resumed at a feverish pace—over two hundred thousand arrivals a year between 1867 and 1869.

Conditions for immigrants on the passage from Europe were often appalling. Shipowners crammed as many people as possible into the holds and provided them with rotting food and foul water. Many of the ships, especially the sailing ships long past their prime, became virtual death ships on which hundreds died on the voyage to the United States. Despite the poor food and awful accommodations, shippers charged between seventeen and twenty-five dollars a person for the passage from Liverpool, England, and Cork, Ireland, to New York. For the Irish, with famine at home, any sacrifice seemed worth the escape to America.

Both the British and American governments passed laws designed to improve the conditions on the immigrant ships, but despite new legislation the dreadful conditions persisted. Far more dangerous than just bad food, crowded conditions, no privacy, and no sanitation, the specter of disease and death stalked the immigrant ships, especially cholera and typhus. During the first four months of 1853, a total

Table 1.4. Immigrant arrivals at Castle Garden, New York, 1855–1869

Year	No. of immigrants
1855	136,233
1856	142,342
1857	183,773
1858	78,589
1859	79,322
1860	105,169
1861	65,589
1862	76,306
1863	156,844
1864	182,296
1865	196,352
1866	238,418
1867	242,731
1868	213,686
1869	258,989
Total	2,356,639

source: Frederick Knapp, *Immigration and the Commissioners of Emigration* (New York: The Nation Press, 1870), 236.

Table 1.5. Number of deaths on selected immigrant ships arriving in New York from Liverpool, 1853

Ship	Arrival	Passage (days)	No. of passengers	No. of deaths
Great Western	Sept. 10	31	832	0
Lucy Thompson	Sept. 12	29	835	40
William Stetson	Sept. 12	31	335	0
Isaac Webb	Sept. 19	29	773	77
Roscius	Sept. 20	35	495	0
Montezuma	Oct. 18	25	404	2
Washington	Oct. 24	41	952	81

SOURCE: Knapp, *Immigration and the Commissioners of Emigration*, 35.

of 96,950 immigrants left Europe on 321 ships bound for New York. Aboard 47 of these ships cholera struck, and 1,933 passengers died at sea—8.5 percent of all those who sailed on the 47 ships.[24] More than half of the cholera ships left from Liverpool, crammed with Irish immigrants, after having stopped in Ireland to load their "steerage" cargo. Deaths varied from ship to ship, even from the same port (table 1.5).

Among those on the *Lucy Thompson*, Mrs. Charles Tierney's ten-year-old son, Charles, died on the voyage, as did Richard Simon, a fifty-six-year-old laborer whose son, Harry, twenty, and daughter, Ellen, seventeen, survived the voyage to New York. Mary Nolan, forty-five, died at sea and left her two sons, Michael, twenty, and Patrick, twenty-five, to fend for themselves in America. The *Lucy Thompson* and *William Stetson* were just two of fifteen ships loaded with immigrants that arrived in New York harbor on September 11, 1853—fifteen ships on just a single day![25] Desperate to leave Ireland, these people crowded aboard ships like the *Lucy Thompson* and risked all for the chance, the opportunity, of a better life in America. Some died before they even caught a glimpse of the golden door which the immigrant ships passed through between Staten Island and Brooklyn, through the Narrows to the Upper Bay and the beacon for all— Manhattan Island.

A Factory World

The one resource the immigrants arrived with was their labor. Economically, immigration enabled America's industrial revolution to expand dramatically, creating insatiable demands for labor. During the nineteenth century, immigrants flooded into the new factories, supplying the labor for New York's industrial revolution. In turn, the factories needed raw materials, and their finished goods had to be transported, creating further need for efficient transportation in New York and its port.

With all of its new workers, New York became the center of the country's expanding manufacturing, at the heart of the American industrial revolution. The transportation revolution continued with the rise of manufacturing, for thousands of tons of raw materials and finished goods had to be moved about the island and across the harbor each day. Freight traffic hauled by the New York Central and the other railroads serving the city and harbor grew exponentially as a consequence.

New York State's census of 1855 detailed the increasing importance of "manufactures" to the state's economy: "The manufactures of the State are very extensive, embracing an almost endless variety of articles. In many sections, the manufacturing interests surpass those of agriculture or commerce."[26] The census enumerated 4,456 manufacturing establishments across the state. In New York City 1,218 manufacturers accounted for 27 percent of the entire state's manufacturing. Brooklyn was home to 221 firms, which included seven distilleries, employing 215 men and producing $2,499,000 worth of liquor, primarily rum. Along the Brooklyn waterfront, ten ropewalks with 677 workers produced rope worth $2,206,153. Sailing ships, still crowding New York harbor, needed miles of rope to control their sails; and hard use demanded constant replacement, creating a lucrative seaport industry.

Just five years later, the 1860 Federal Census of Manufactures reported 4,375 establishments in New York City—four times as many as in 1855—with 90,204 employees producing products valued at $159,107,369. Garment, boot, and shoe firms accounted for the major share of the city's manufacturing. Across the East River in Brooklyn, 1,032 firms—a fivefold increase—reported a total output of just over $34 million, led by sugar refining and liquor distilleries.[27] Together, New York and Brooklyn manufacturing accounted for 10 percent of all manufacturing in the entire country.

The Civil War created enormous demand for manufactured goods, and New York's business expanded to meet that demand. The ship yards along the East River built hundreds of ships for the federal navy, including the North's first ironclad ship, the *Monitor*, constructed at the Continental Iron Works on the East River in the Greenpoint section of Brooklyn. New York's expanding garment industry made hundreds of thousands of uniforms for the Union army.

Industry in the port of New York continued to grow in the decades after the Civil War. The 1880 Report on Manufactures provided a detailed analysis of the state of manufacturing. New York dominated in numerous categories, and if Brooklyn is added, the central role of the city in the nation's manufacturing only increases. A total of 16,530 manufacturing companies employed a workforce of 183,284 men, 78,815 women, and 12,835 "children and youths."[28] A more telling measure of New York's ascendancy was the city's share of the value of manufactured goods: New York and Brooklyn firms accounted for 12.2 percent of the value of all manufacturing in the entire country.

Tables included in the 1880 Report on Manufactures ranked the seven leading cities in the country by industry and aggregate production. Of the thirty-seven industries listed, New York ranked first for eleven, and Brooklyn was first for two

Table 1.6. Leading industries in New York City and Brooklyn by national ranking in production, 1880

National ranking	City	Industries
1st place	New York	Blacksmithing, bread and bakery products, carpentering, men's clothing, women's clothing, foundry and machine shop, furniture, liquor's malt, marble and stonework, printing and publishing, tobacco and cigars
	Brooklyn	Cooperage, sugar and molasses
2nd place	New York	Carriages and wagons, planed lumber, mixed textiles, shipbuilding, silk and silk goods, slaughtering and meatpacking, tinware and copperware
	Brooklyn	Drugs and chemicals

SOURCE: U.S. Census Bureau, *Report on the Manufactures of the United States at the Tenth Census, June 1, 1880* (Washington, DC: Government Printing Office, 1883), xxvi.

more; together they ranked first or second for twenty-one of thirty industries (table 1.6).

The sheer diversity of manufacturing was stunning. Shipbuilding had been an important part of the port's economy since the Dutch arrived in the 1600s and remained so. New York now led the country in clothing manufacturing as well as foundry and machine-shop products. The city's machine shops produced the steam engines that replaced sail for American-built ships and also exported steam engines to other countries, competing with British manufacturers. During the Civil War, the Delamater Iron Works on 13th Street, near the East River in Manhattan, constructed the engine for the *Monitor*, the ironclad vessel which fought the *Merrimack* in the Chesapeake Bay on March 9, 1862.

The 287 foundry and machine-shops on Manhattan Island employed almost ten thousand workers. Burrows and Wallace, the Pulitzer Prize–winning authors of *Gotham*, use the term "Iron Age" to characterize the rise of the heavy metal industry in the city. Coal and iron ore hauled by the railroads from Pennsylvania or by barge over the Delaware & Hudson Canal had to be moved across the Hudson and up the East River, to feed the furnaces at the famous Novelty Iron Works at 12th Street. The Morgan Iron Works occupied two full blocks between 8th and 10th streets from Avenue C to the East River; it had three piers for barges to unload coal and iron ore. The cast-iron facades produced by Architectural Iron Works, on 14th Street and Avenue C, for New York's new shopping "palaces" and thousands of storefronts and office buildings created the city's "cast-iron" districts. Novelty Iron Works and the Morgan Iron Works each employed as many as a thousand workers, many of them immigrants who lived in the neighborhood. Smoke and coal dust filled the air as the furnaces ran twenty-four hours a day all year long.

All of the manufacturing in New York and Brooklyn created enormous transportation demands. Raw materials had to be transported by rail or barges and ferries across the city's water boundaries to the Manhattan or Brooklyn waterfronts and then hauled to the thousands of individual companies. For finished products that were not destined for local consumption, the process had to be reversed. Horse-drawn carts, or drays, filled the city streets and crowded the piers lining the waterfront. On the piers an army of stevedores and laborers loaded goods from the drays onto ships and barges. Tens of thousands of horses provided the muscle power to haul goods on the city's streets day and night and in all weather throughout the year.

Transportation Vortex

By the time of the 1880 census, New York dominated the country's shipping, commerce, and manufacturing. It continued to serve as the main entry port for the millions of immigrants flooding to the shores of America. The city's population ranked first in the nation, with 1,206,299 residents. Brooklyn ranked third with 586,663 residents. Combined, the population of Manhattan and Brooklyn totaled more than twice the population of the country's second largest city, Philadelphia (population 847,170). Chicago, the rising colossus of the Midwest, had a much smaller population of just over 500,000.[29]

New York's preeminent role in the nation's import and export trade attracted a greater and greater share of the world's shipping to the port. By the turn of the twentieth century, at the time William Wilgus arrived to work in New York, over half of the country's imports and exports passed through the port. Thousands of ships passed through the port of New York each year; often, hundreds of ships arrived and departed each day. Outbound ships departed not just for Europe, but for the entire world. Coastal vessels traveled to and from New York to southern ports and through Long Island Sound to New England.

No other port in the world, not even London or Rotterdam, equaled New York in the sheer volume of shipping. In 1900 a total of 4,348 ships arrived from foreign ports, the majority of them steamships but also a few sailing vessels.[30] The U.S. Customs Service carefully monitored all shipping to and from foreign ports in order to collect customs duties. Sailing vessels, particularly schooners, still dominated coastal shipping. The *Commercial Year Book* reported a total of 6,359 coastal ship arrivals from southern ports, of which 4,475 were schooners and 1,808 were steamships. From New England, 4,662 ships arrived in New York, most of them wind-powered schooners.[31] Some of the sailing vessels also had steam engines aboard to provide auxiliary power when the wind died. Official statistics totaled the number of ships arriving in 1900, in both foreign and coastal trade, at 15,496—an immense fleet filling the harbor with over 500 ships a day.

It is hard to imagine today the incredibly crowded conditions in the port at the turn of the twentieth century. The 1892 "bird's-eye" lithograph of lower Manhattan shown in the introduction to this book captures some of the crowded shipping activity on the Hudson and East Rivers. Not drawn to scale, the print depicts hundreds of river steamers, ocean steamships, sailing vessels, ferries, and tugboats crowding the water and jamming the piers along the rivers and on the New Jersey shorefront in the foreground. A more crowded maritime scene could not have been depicted anywhere else in the world.

New York's dominance of the nation's shipping, well established before the Civil War, became self-reinforcing. Not only American shipping companies, but also foreign shippers located their North American operations in New York and sent their ships to and from the port and not from Boston, Baltimore, or Philadelphia. In 1900, there were seventy-six steamship lines operating between New York and foreign ports in Europe and around the world. Among the sixteen lines with service to Great Britain, the famous Cunard and White Star Lines offered luxurious passenger service. The more modest Bristol City Line carried only freight every five days to Bristol and other ports along the English Channel.

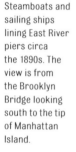

Steamboats and sailing ships lining East River piers circa the 1890s. The view is from the Brooklyn Bridge looking south to the tip of Manhattan Island.

Hamburg-American Line and North German Lloyd both ran their ships from New York to Hamburg and Bremen Germany. North German Lloyd advertised a special express service every Thursday, with an average voyage of just six and a half days, to Hamburg with stops in Plymouth, England, and Cherbourg, France. Employing a number of steamships in continuous service, including its brand new *Deutschland*, the firm's ships connected in Hamburg to Africa and Asia. Nine lines provided service to the Mediterranean and five to France, including France's premier shipping company, Compagnie General Transatlantique.

Sixteen firms offered service to India, China, Australia, and the Far East.

Table 1.7. Ships and tonnage through the port of New York, direct foreign service, 1907–1919

Year	Entered harbor		Cleared harbor	
	No. of ships	Tonnage	No. of ships	Tonnage
1907	4,315	11,984,297	3,863	11,473,334
1910	4,419	12,405,673	3,755	12,962,709
1913	4,448	13,893,249	4,203	15,157,301
1916	5,192	13,213,919	5,388	13,843,473
1919	5,016	13,974,052	5,002	14,427,029

SOURCE: New York, New Jersey Port and Harbor Development Commission, *Joint Report with Comprehensive Plan and Recommendations* (Albany, NY: J. B. Lyon, 1920), 175.

These included the Anchor Line to Bombay and Calcutta, and Peabody's Line to Australia and New Zealand via South Africa. The U.S. China-Japan Steamship Line ships sailed by way of the Mediterranean and the Suez Canal to Singapore, Hong Kong, Shanghai, and Kobe and Yokohama, Japan. Twenty-five different steamship companies offered passenger and freight service to South and Central America, Mexico, the Caribbean, and Bermuda.[32] From New York freight could be shipped to every major port in the world. New York's rival ports—Boston, Baltimore, and Philadelphia—offered service to foreign countries, but not one of them came close to the breadth of worldwide destinations emanating from the piers lining the Hudson and East Rivers and throughout the port of New York.

New York's central place in the nation's commerce continued unchallenged for the next decades. At the end of World War I, at the urging of William Wilgus, the states of New York and New Jersey finally began joint planning for the future of the port by establishing the New York, New Jersey Port and Harbor Development Commission in 1917. The Port Commission, the precursor of the Port Authority, conducted an exhaustive study of the port and its transportation needs. As part of the study, the commission assembled detailed information on ships entering and leaving ("clearing") the port for the years 1907 through 1919.

The commission report broke out data for ships engaged in "foreign direct service," from New York to foreign ports. Measured either by the number of ships in foreign trade or the tonnage shipped through the port, international commerce constituted a major share of the city's business activity. Foreign shipping, in 1919, generated over ten thousand ship movements to and from the harbor, loaded with over twenty-eight million tons of cargo (table 1.7). Estimating the tonnage carried in coastal trade proved more difficult, but in 1914 the report listed the number of ships cleared "coastwise in bond" (excluding Hudson River and Long Island Sound vessels) at 2,482, carrying nearly six million tons of cargo.[33] The Merchants' Association publication *Greater New York* summarized: "The supremacy of the Port of New York in foreign commerce is once more indicated."[34]

Tonnage numbers arriving in the port provide only one indicator of the incredible complexity of the activity generated. All of the cargo arriving in New York to be loaded for transport to distant foreign shores, to be sent along the coast to Savannah or north to Providence, or to be distributed locally in Manhattan or Brooklyn required complex handling. New York harbor became the center of a huge transportation vortex, drawing an ever-increasing volume of freight and passengers. The thousands of ships, hundreds of passenger trains, and thousands of freight cars crowding the harbor and rail lines created unimaginable complexity.

Arriving cargo had to be unloaded from ship to pier or to a "lighter," a barge moored next to the ship. Once the cargo was on the pier, men with hand trucks moved it to an assembly area inside the pier shed to be sorted. With all of the necessary paperwork completed, hand trucks moved the goods to a loading dock, where horse-drawn drays, or by 1914 trucks, waited to move the cargo to its destination. Some of the inbound cargo moved along crowded city streets to other piers to be loaded on another ship for transport along the coast. A significant part of the total tonnage consisted of freight bound for the city's thousands of manufacturing companies and food to supply the city's population.

Since almost half of all of the entire country's imports and exports moved through the port of New York, a major share of cargo required transportation by rail to and from other parts of the country. If a ship berthed along the Hudson River, the New York Central's freight lines down the west side of Manhattan provided easy rail access. The Central's tracks did not, however, extend onto the piers; most freight destined for the Central had to be carted a couple of blocks inland to freight depots and then loaded onto freight cars.

Legions of longshoremen labored on the city's waterfront and in freight depots, providing the raw muscle power needed before freight containers and huge gantry cranes replaced the small army of men who swarmed over the ships, piers, and warehouses. Of course, all this labor needed to move freight back and forth created significant costs. In 1907, Wilgus, advocating for a freight subway to connect New Jersey and Manhattan, estimated freight handing in the port at over $2.50 a ton. To put this figure in perspective, the cost of hauling a ton of freight the nine hundred miles from Chicago to New York averaged $40 a ton; handling in New York thus amounted to over 5 percent of total transportation cost.

In 1919 the New York, New Jersey Harbor Development Commission studied, in great detail, the labor and costs involved in loading and unloading ships. Researchers selected nine ships representing different types of trade and carried out meticulous "clockings" of all cargo transfer operations, examining four separate activities: "breaking out," "hoisting," "trucking," and "stowing" for inbound cargo and the reverse for outgoing cargo. To unload 2,537 tons of cargo from a large transatlantic steamer at a Brooklyn pier required 4,263 man-hours of labor and took three full days, with 20 percent of the labor involving overtime. Unloading freight from the ship to the pier warehouse cost $1.28 per ton.[35] Additional charges were incurred in moving the freight from the pier to its final destination. The costs for the eight other ships in the study varied depending upon

the ship's cargo-handing configuration and the type of cargo. Costs also varied from pier to pier depending on how modern the pier's cargo handling facilities were.

Freight destined for the massive rail yards on the New Jersey side of the harbor where eight major railroads maintained facilities required additional effort. Hundreds of tons of crates or barrels filled with freight or sacks of raw materials stowed in the holds of arriving ships had to be unloaded and then floated across the Hudson River to be reloaded onto waiting freight cars. Exports required the process to be reversed.

Each of the railroads serving New York maintained a small fleet of car floats, lighters, barges, cattle boats, ferries, and tugboats. The Pennsylvania Railroad operated 27 passenger and freight steamboats, 23 ferryboats, 124 car floats, 9 lighters with steam cranes, 226 barges, and 55 tugboats. The New Haven Railroad ran a more modest maritime fleet with a single ferry, 18 tugs, and 54 car floats, while the Baltimore & Ohio operated 13 tugboats, 49 car floats, 3 steam lighters, 92 lighters, and 3 steam-derrick lighters.[36] All of the other major railroads serving the port of New York (Erie; Lehigh Valley; Central of New Jersey; Long Island; Philadelphia & Reading; West Shore; New York, Ontario, & Western; Delaware, Lackawanna & Western) operated similar marine equipment.

Even the New York Central, which had the enormous advantage over its competitors of rail lines on Manhattan Island, maintained a fleet of 10 ferry boats, 21 tugs, 63 car floats, and 174 barges of various sorts. The Central's equipment transported freight back and forth across the Hudson to the West Shore railroad facilities at Weehawken, New Jersey. The Central leased and operated the West Shore Railroad as an integral part of its extensive railroad system.

Thus, in addition to ocean-going and coastal ships, hundreds of tugboats, car floats, barges and lighters crowded the harbor each day hauling tons and tons of freight. Nighttime brought only a slight halt to all of this activity. Skilled captains carefully navigated the harbor to avoid collisions. Whenever fog rolled in, ships sounded their foghorns every few minutes in an attempt to establish their position and define their right-of-way. Collisions occurred on a regular basis, some with loss of life and serious damage to both vessel and cargo. When the weather was especially cold, ice created further hazards; and on rare occasions, the harbor froze over, bringing not only shipping but also the entire city to a standstill.

To add to the chaos created by harbor commerce, numerous passenger ferries provided service for passengers and horse-drawn drays and trucks across the East River, the Hudson River, and to Staten Island. Sixteen ferry routes connected Manhattan and New Jersey in 1914; the farthest north ran between Edgewater, New Jersey, and 125th Street. A number of the major railroads that terminated in New Jersey provided ferry service across the harbor for their passengers and the general public. For many years the Pennsylvania Railroad operated the Pavonia Ferry between Jersey City and Chambers Street in lower Manhattan. To this day many Staten Island residents ride the Staten Island Ferry daily to and from Manhattan following the very route Commodore Vanderbilt's first boat sailed in 1810. On the East River at the turn of the twentieth century, sixteen ferries crossed the river, including the usually crowded and famous Fulton Street Ferry. Of course, ferry service competed with shipping for precious pier space, especially on the Manhattan waterfront. Once established, the ferry slips could not be used for any other purpose. On the very busiest ferry routes, two boats usually operated simultaneously with two ferry slips on each shore.

All Railroads Drawn to New York

For the Vanderbilts and their New York Central railroad empire, direct rail access to Manhattan Island for both passengers and freight proved to be a key to unimaginable riches. While the glamorous Grand Central Terminal received rave reviews and served as a fitting symbol for the second largest railroad company in the country, for the railroad's bottom line its freight lines were more lucrative. Along the west side of Manhattan, close to the piers lining the Hudson River, workaday steam engines hauled freight cars to and from the piers and warehouses and earned millions for the corporate coffers.

New York Central's competitors did not stand idly by and watch the railroad reap all of the advantages from the port's pivotal role as the center of nation's import and export trade and as a manufacturing center. Over time, all of the major east coast railroads fought to bring their tracks to the shores of the port. With the one exception of the New York Central, their tracks stopped at the edge of the harbor on the New Jersey side. A mile of water separated the railroads from the tantalizing prize in easy view just across the river.

As all the rival ports fell farther and farther behind New York, even the railroad companies first established to serve the ports of Boston, Baltimore, and Philadelphia turned their energy to establishing rail links to the port of New York. Even if their lines were inconveniently located on the Jersey side of the harbor and very expensive to operate, none of the railroads believed they could prosper without offering service to New York. Eventually all of the major railroads serving the East Coast arrived in the port, creating a virtual maze of railroad tracks, train yards, freight facilities, and passenger terminals up and down the length of the New Jersey shore opposite Manhattan and farther down the harbor all the way to Staten Island.

Each railroad's history differed. The Erie Railroad, chartered by the state of New York in 1832, constructed its original rail line from Piermont, New York, on the Hudson River, across the southern tier of New York, to Dunkirk on Lake Erie south of Buffalo, a distance of 450 miles. When completed in 1851, the Erie formed the longest railroad in the world. Piermont, twenty miles north of New York on the west side of the Hudson, seemed an illogical terminus for the first major trunk railroad in the country—and an ambitious railroad at that. The Erie's charter from the state required that all the railroad's tracks be within state boundaries and Piermont is just north of the New Jersey border. Ferries hauled passengers and freight from Piermont down the Hudson to piers at Duane Street in Lower Manhattan, hardly convenient and of course very costly.

The Erie executives knew they needed more direct access to New York City. A number of small, New Jersey–based railroads had already built rail lines from Jersey City north to the border of New York State in an effort to attract some of the Erie's business. A bitter battle ensued with the Erie; and in 1852 the Erie acquired, by lease, all of the small railroads, creating direct rail links to Jersey City, where the Erie opened a large passenger terminal, a ferry landing, and a freight yard in 1862. Now Erie passengers and freight could travel from the shore of Lake Erie to the shore of the Hudson—but still faced an additional ferry ride across the Hudson River, the all-important "last mile" to arrive in Manhattan.

The Erie's extensive rail facilities in New Jersey included two major train yards in Jersey City, one along the waterfront and the second in the Croxton neighborhood along the Hackensack River. Each yard contained miles of tracks to sort cars, handle huge amounts of freight, disassemble inbound trains, and assemble outbound trains. Hundreds of steam engines and thousands of employees labored night and day. Across the river in Manhattan, the Erie occupied piers 20, 21, and 39 on the Hudson River and pier 7 on the East River, with large warehouse facilities to store freight. At 28th Street, the railroad operated a "float-bridge" terminal where car floats docked and steam engines unloaded freight cars directly onto railroad tracks on the city streets. In Jersey City the Erie also had a grain elevator, a milk station, and stock yards for live cattle shipped from the Midwest. For a long time consumers preferred to have beef freshly slaughtered rather than eating "dressed" beef from the Chicago slaughterhouses. With all of these facilities in New Jersey, the railroad's harbor "fleet" included 11 ferry boats, 15 tugs, 36 car floats, 6 steam barges, 138 covered barges, and 60 open scows.[37] Both the geography of the port of New York and the Central's monopoly on direct rail access to Manhattan forced the Erie to establish this costly infrastructure.

Like the Erie, the mighty Pennsylvania waged a long battle to gain access to the New York market for both freight and passengers. The Pennsylvania, at enormous expense, managed to bring its passenger service onto Manhattan Island. Willing to spend millions of dollars, the Pennsylvania first gained control of the

Long Island Railroad in 1900. Then, with its rail lines already to the New Jersey side of the Hudson River, the railroad obtained state and federal permission to build tunnels under both the Hudson and East River. The railroad quietly purchased two complete blocks of Manhattan real estate at 33rd Street and, while constructing the tunnels, began to build a monumental passenger station. Pennsylvania Station, a Beaux-Arts masterpiece designed by McKim, Mead & White, opened with great fanfare in 1911, almost two full years before the new Grand Central Terminal opened at 42nd Street.[38]

While the Pennsylvania Station garnered wide attention in the press and praise from the traveling public, the Pennsylvania Railroad's main source of profit, as with all the railroads, remained the freight business. And despite spending over $100 million dollars on the Pennsylvania Station project, the railroad's freight lines still terminated at Jersey City across the Hudson River from lower Manhattan. The Pennsylvania's tunnels and magnificent station at 33rd Street served only passenger trains. The railroad could not offer the same direct service to its freight customers.

The Pennsylvania operated major rail yards in Newark, in the Meadowlands west of Jersey City, and in Jersey City itself. To bring freight across the harbor, it maintained extensive facilities on Manhattan Island, in Williamsburg, and at Bay Ridge Brooklyn. Bay Ridge, further to the south of the Brooklyn waterfront, served as a terminus for car floats linked to the Pennsylvania tracks at Greenville, on the Jersey side of the harbor. At Bay Ridge a freight line connected to the Long Island Railroad's yard at Sunnyside, Queens.

In an attempt to outflank the New York Central, the Pennsylvania and New Haven railroads formed the New York Connecting Railroad to link the Long Island Railroad tracks at Sunnyside to the New Haven through the Bronx to New Rochelle in Westchester County. To accomplish this task the railroads spent a massive amount of money to construct the Hell Gate Bridge in 1917, crossing the East River between Queens and the Bronx. With the completion of the bridge, the Pennsylvania floated freight cars across the harbor to Bay Ridge, then on to the Sunnyside freight yard. From Sunnyside the New York Connecting Railroad crossed over the Hell Gate Bridge, through the Bronx to join the New Haven's line at New Rochelle. Convoluted to the extreme, this complicated rail link enabled the Pennsylvania to offer freight service from the Midwest to Long Island and New England despite all the time and expense. It remains unclear if the railroad ever made money hauling freight to either Long Island or New England via Bay Ridge and the Hell Gate. To this very day there is still no direct freight service across New York harbor.

In addition to its vast rail yards in New Jersey, the Pennsylvania leased eight piers on the Hudson River—3, 4, 5, 27, 28, 29, 77, and 79—and pier 22 on the East River. To cross the harbor the railroad operated a virtual "navy" to provide freight service around the port of New York: 27 passenger and freight steamboats, 55 tugboats, 124 car floats, 226 barges, and a motley collection of other barges and scows. All of this maritime equipment cost a small fortune to acquire and operate.

All of the other railroads terminating on the Jersey side of the harbor—the

Table 1.8. Freight tonnage on Hudson River railroad piers, 1914

| Rail Line | Tonnage | No. of railcars | | |
		Inbound	Outbound	Total
New Jersey				
Ontario & Western	52,567			
Lehigh Valley	342,313	23,794	23,794	47,588
Baltimore & Ohio	406,266	25,357	25,357	50,714
West Shore	451,940	40,681	41,271	81,888
Lackawanna	476,943	38,593	38,593	77,186
Central of New Jersey	607,413	48,556	48,556	97,112
Erie	663,666	58,478	58,487	116,974
Pennsylvania	1,478,004	113,156	113,156	226,312
Subtotal	4,479,112	348,615	349,214	697,774
New York Central	1,672,445	135,754	135,234	270,988
Grand Total	6,151,557	484,369	484,448	968,762

SOURCE: Port and Harbor Development Commission, *Joint Report*, 151.

Baltimore & Ohio (B&O), Lehigh Valley, Philadelphia & Reading, and Delaware, Lackawanna & Western—maintained similar facilities and equipment, although not on the same scale as the Pennsylvania Railroad. In addition to their fleets of ferries, tugboats, car floats, and barges they employed a small army of railroad workers to operate the maritime equipment day and night throughout the year, adding to the overall cost of handling freight. These costs only magnified the advantage enjoyed by the New York Central & Hudson River Railroad because of its direct freight service onto Manhattan Island.

In order to compete with the Central, none of the other railroads serving the port of New York charged an additional fee for ferrying freight across the harbor. The railroads hauled freight to and from New York for the same price as the Central, with its significantly lower operating costs. A long-running political and legal battle ensued over freight rates to be charged. Each railroad fought to gain as large a share of the traffic as possible, even with the significant cost associated with the "last mile" of service from New Jersey to New York. Businesses and shippers across the country demanded service to Manhattan and Brooklyn, not just to the Jersey shore of the harbor.

On the other side of the harbor New Jersey politicians and business leaders argued loudly that if the railroads charged fees reflecting the true costs of delivering or receiving freight in Manhattan or Brooklyn, the freight business would move across the harbor to the Jersey shore. New Jersey would benefit enormously, gaining thousands of waterfront jobs and investment. They envisioned miles of new piers lining the New Jersey side of the harbor with hundreds of ships berthing each day.

Table 1.9. Hudson River pier stations, 1914

Pier	Cross street	Railroad
3	Morris	Pennsylvania
4	Morris	Pennsylvania
5	Morris	Pennsylvania
8	Rector	Lehigh Valley
10	Albany	Central of New Jersey
11	Cedar	Central of New Jersey
13	Fulton	Delaware, Lackawanna
16	Barclay	New York Central
17	Park Place	New York Central
20	Chambers	Erie
21	Duane	Erie
22	Jay	Baltimore & Ohio
23	Harrison	West Shore
27	Hubert	Pennsylvania
28	Laight	Pennsylvania
29	Vestry	Pennsylvania
31	Desbrosses	New York Central
34	Canal	Lehigh Valley
39	Houston	Central of New Jersey
41	Leroy	Delaware, Lackawanna
46	Charles	Central of New Jersey
66	West 27th	Lehigh Valley
68	West 28th	Delaware, Lackawanna
77	West 37th	Pennsylvania
78	West 38th	Pennsylvania
81	West 41st	Central of New Jersey
83	West 43rd	New York Central

SOURCE: Port and Harbor Development Commission, *Joint Report*, 134.

But hauling freight to New York provided the railroads on the New Jersey side of the harbor with a huge volume of shipping and justified their efforts over time to bring their rail lines to the port. No matter how expensive service to New York was, no major railroad serving the East Coast could survive without it. All of the major railroads had to "come to New York," despite the cost. Maintaining a small fleet of vessels to provide the last mile of service seemed to be a small cost to pay for access to the most important manufacturing and shipping center in the world.

In 1914, the Hudson River piers occupied by the railroads alone handled over six million tons of cargo, requiring the movement of 968,762 freight cars (table 1.8). Even with the Central's decided advantage, the eight New Jersey railroads handled a larger volume of freight; a number of them provided service to parts of the country the New York Central did not.

To move this enormous amount of tonnage to and from Manhattan the various railroads occupied twenty-seven piers on the Hudson River and seventeen on the East River (table 1.9). Commerce completely monopolized the shores of Manhattan Island and the Brooklyn waterfront. From the Battery and up the Hudson and East Rivers for mile after mile piers crowded the waterfront, eliminating any other activity and cutting New Yorkers off from the waters surrounding Manhattan. The occupation of all of this valuable pier space by the railroads created further chaos in the harbor. Even though hundreds of ships were arriving each day, often no pier space was available, and ships had to anchor in the Upper Bay awaiting their turn to berth when space at a pier opened.

Many of the freight cars, once unloaded, had to be hauled away empty. The Central's freight cars crowded its tracks on the west side of Manhattan day and night, creating the "West Side problem" William Wilgus proposed to solve with a freight subway and tunnels under the Hudson River linked to the New Jersey railroads' freight yards. Freight bound for Manhattan would be transferred to the subway, eliminating the need for railroad piers on the Hudson and dramatically reducing the fleet of tugs, barges, and car floats that continually crossed the harbor.

No part of this huge transportation system resulted from a rational, planned process. Over three hundred years, piecemeal development had created the most complex ship, water, river, pier, and rail interchange in the world, driven by the vested interests of thirteen major railroads, hundreds of shipping companies, the cities of New York and Brooklyn, and the numerous municipalities across the harbor in New Jersey. Add the conflicting priorities of the states of New York and New Jersey, and a permanent state of chaos ensued. By the time Wilgus arrived in New York in 1899, the size and scale of the port of New York and its complex haulage system seemed beyond rational control. He devoted his career to an attempt to bring some order to this chaotic transportation system in the metropolitan region—a daunting task and one that eludes us to the present day.

2

THE BRILLIANCE
OF GRAND
CENTRAL

WILLIAM WILGUS ACHIEVED meteoric success with the New York Central
Railroad. In June of 1893 he joined the
Rome, Watertown & Ogdensburg Railroad in upstate New York, a subsidiary
of the Central. Wilgus reorganized the
company's engineering records and un-
dertook a careful examination of the tracks, bridges, and culverts in need of immediate repair.
His hard work impressed Charles Russell, superintendent of the Rome, Watertown & Ogdensburg; and he came to the attention of Colonel Walter Katte, the chief engineer of the entire New
York Central system based in New York City.

As his work responsibilities expanded, Wilgus achieved a measure of professional stature,
always important to his sense of self. In 1896, the American Society of Civil Engineers elected
him to full membership at thirty years of age, the earliest age possible. He proudly remained a
member of the ASCE for the rest of his life.

His first opportunity to work directly for the senior management of the Central involved an inspection of the Ogdensburg & Lake Champlain Railroad, a
railroad running from Ogdensburg to Lake Champlain, soon to be purchased by
the Central.[1] After just four years in upstate New York, he received an offer of the
position of resident engineer of the eastern division of the New York Central. He
immediately accepted and moved to New York City to work out of the corporate
offices at Grand Central Depot at 42nd Street. Just thirty-four years old, William
Wilgus now held a senior engineering position with the second largest railroad
in the United States, a singular achievement for a young, self-taught engineer.

Wilgus's first major assignment involved the planning for the Terminal Railway of Buffalo, which was to bypass the congested rail lines in downtown Buffalo
and provide a speedy link for trains running from the Midwest on the main line
of the New York Central across upstate New York. With the successful completion of Terminal Railway, in just six months, another promotion awaited: chief
assistant engineer of the entire New York Central system. In his autobiography
Wilgus notes that a careful inspection revealed the main line tracks across upstate New York to be "seriously suffering from dry rot. . . . the bones of the corpus
were decaying," conditions that were, in his opinion, "out of sight [and] out of
mind to those aloft who were blind to what was going on beneath."[2]

Further advancement followed, and in 1898, the new president of the Central,

William H. Newman, moved Wilgus into the newly created position of chief of maintenance of right-of-way, responsible for all of the system's tracks. He set out to professionalize his department, replacing older railroad veterans with younger, professionally trained engineers. Wilgus did not hesitate to force out long-serving employees, "bringing sorrow to many who had been long in the service." His goal: to hire men who would become "pioneers in the newly established organization in which engineering talent has a wider sway than of old." He assembled a top-notch engineering department to overhaul the railroad and, in the not-too-distant future, plan for a new Grand Central. Wilgus's initiatives pleased senior management, including the grandsons of Commodore Vanderbilt, William K. and Cornelius III, who remained very active in all major decision making. In his autobiography Wilgus commented on his increasing responsibility and the costs involved: "For the first time in my life I was learning the truth of the old saying—uneasy rests the head that wears the crown."[3]

Just a year later on April 15, 1899, President Newman promoted Wilgus to chief engineer of the entire New York Central & Hudson River Railroad. In the space of six years with the New York Central, he had advanced from a minor position with a subsidiary in Watertown, New York, to the very heart of the railroad, working at corporate headquarters at Grand Central Depot in midtown Manhattan.

Challenges in New York

Wilgus recognized that the Central's major challenges in New York City involved both the railroad's passenger and freight service. A never-ending conflict pitted local residents, politicians, and the general public against the railroad. Grand Central Depot, the railroad's passenger terminal at 42nd Street, which had opened with such fanfare in 1871, had reached capacity and could no longer handle the volume of passengers. On the west side of Manhattan, the railroad's extensive freight tracks and rail yards along the Hudson River, extending all the way down to the tip of Manhattan Island, created a maze of tracks, train yards, and storage depots that disrupted the city's streets. While the Central viewed its right to use the city streets as absolute, based on the charter granted to the Hudson River Railroad fifty years earlier, the public was increasingly demanding that the railroad solve the "West Side problem" created by its freight facilities along the Hudson. Commercial, railroad, and shipping activity dominated the entire Hudson River waterfront below 70th Street. As a result of over more than two hundred years of development, the waterfront was a commercial and industrial landscape far removed from the lives of the residents of the island.

Cornelius Vanderbilt's Grand Central Depot at 42nd Street no longer served the railroad's or the traveling public's interests. In thirty years the number of passengers using the depot had increased beyond its builders' wildest projections. More and more long-distance trains arrived and departed each day; the number of suburban commuters also had increased dramatically. By the time Wilgus arrived in New York, Grand Central Depot served over 15 million passengers a year, averaging 48,600 a day.[4] With only fifteen tracks in the train shed, the depot barely accommodated the number of scheduled passenger trains each day. New Yorkers felt the outdated facility no longer reflected the dynamism and energy of the most important city in the United States. A *New York Times* editorial harshly criticized the depot: "Nothing pertaining to New York City except its government has been so discreditable to it as its principal railroad station. . . . [It is] wretchedly cramped in space . . . ill-arranged, dark, and repelling. . . . Altogether, the Grand Central Station furnished terminal accommodations which would be considered adequate in Sandusky, Ohio."[5] Later in 1899, the *Times* continued to complain: "It is known to travelers as one of the most inconvenient and unpleasant railroad stations in the whole country. . . . The ugly structure has been a cruel disgrace to the metropolis and its inhabitants."[6] The *Times* and the public objected to the lack of grandeur conveyed by Commodore Vanderbilt's old depot. As New York became the center of the country's commerce, many critics believed its principal passenger station should reflect the city's role as the nation's chief port, leading industrial location, center of the financial industry, and the fastest-growing and most densely populated city in the country.

Wilgus also recognized the dangers created by the tunnel under Park Avenue, north of the train yard, stretching from 56th to 79th Streets, through which ran all of the passenger trains to and from Grand Central. Originally the tracks of the New York & Harlem Railroad ran at ground level up the middle of 4th Avenue (Park Avenue) to Harlem. When construction of the railroad began in the 1830s, Manhattan Island at midtown and to the north remained undeveloped open country and farms, but with increasing population, development moved steadily north up the island. First the area around 42nd Street, and then north along Park Avenue and east to the East River, became filled with tenements, factories, and commercial activity.

As more and more people moved to the area around Grand Central Depot and Park Avenue to the north, they objected to the danger, noise, and pollution created by the railroad's trains running down the middle of the street. After tortuous negotiations with the city, the railroad agreed to dig an open cut down the middle of Park Avenue and place the tracks in the bottom of the cut, where the trains could run at faster speeds, separated from the pedestrians and street traffic above. The open cut provided only temporary relief. As train traffic to and from Grand Central increased, calls came to "roof over" the open cut and create a tunnel to further shield the neighbors from the noise, smoke, and soot from the passing trains and their fearsome steam engines. While the public may have gained some relief, the "Park Avenue tunnel" created by the roof trapped all of the heat, smoke, and steam underground, with only a few vents for the steam to escape and allow clean air to enter. Passengers traveling in the tunnel to Grand Central passed through clouds of steam and heat. When inevitable delays occurred, passengers trapped in the tunnel were forced to endure hellish conditions, especially in the summer, until their train could proceed.

A final tragic event sealed the fate of the Commodore's depot and prompted

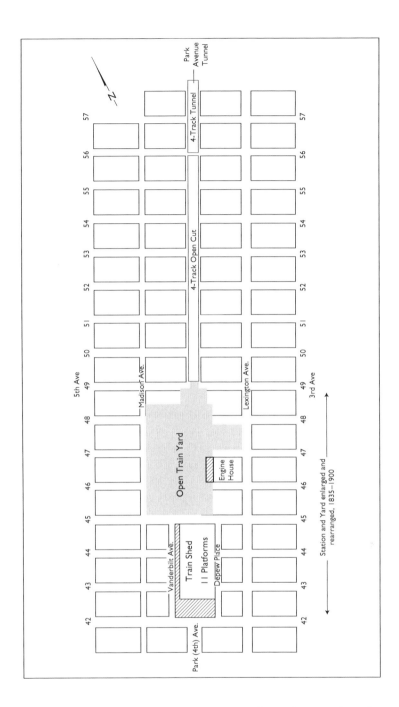

57   57

56   56

55   55

54   54

53   53

52   52

51   51

50   50

49   49

48   48

47   47

46   46

45   45

44   44

43   43

42   42

Park Avenue Tunnel

4-Track Tunnel

4-Track Open Cut

5th Ave

Madison Ave.

Lexington Ave.

3rd Ave

Open Train Yard

Engine House

Station and Yard enlarged and rearranged, 1835–1900

Vanderbilt Ave.

Train Shed

11 Platforms

Depew Place

Park (4th) Ave.

Grand Central depot area, 1900, bounded by Madison Avenue to the west, Lexington Avenue to the east, 42nd Street to the south, and 57th Street to the north.

William Wilgus to visualize an entire new terminal complex to replace the train yard and aging depot at 42nd Street. On January 8, 1902, a morning commuter train from Danbury, Connecticut, stopped in the Park Avenue tunnel to await a signal to proceed further south. With the tunnel choked with steam and smoke, another commuter train from New Rochelle plowed into the rear of the Danbury train. The horrible collision shattered a number of the wooden railroad cars. Coals from the pot-bellied stoves used to warm the passenger cars set the splintered wreckage afire. In the ensuing carnage, fifteen people on the Danbury train lost their lives and hundreds more received serious injuries. The *New York Times* reported that the force of the collision propelled the boiler of the steam engine into the rear car of the Danbury train: "In this small space and in the area above and on the side of the intensely hot boiler were jammed the bodies of sixty passengers who had occupied the car. Men and women were piled in inextricable confusion. Dead bodies in several instances pinned to the floor wounded women. The dead were horribly mangled."[7] Heroic efforts by New York City firemen saved many lives.

An immediate uproar ensued, with calls in the press for the district attorney to indict the engineer of the New Rochelle train and the senior officers of the New York Central for manslaughter. Investigations by the district attorney, the city coroner's office, and the New York State Board of Railroad Commissioners began immediately. While the officers and board members of the Central, including William K. Vanderbilt, expressed their deep sorrow, they all recognized the grave threat the investigations represented. Above all, the New York Central risked losing its incredibly valuable franchise, the right to direct rail access to Manhattan Island for the railroad's passenger trains. Some politicians and members of the press demanded that the railroad move its passenger terminal and steam trains off the island to the Bronx. The Central owned considerable property in Mott Haven, just across the Harlem River in the Bronx. With a relocated terminal in Mott Haven, passengers would transfer to the 2nd or 3rd Avenue elevated railroads to complete their journey to Manhattan.

Wilgus did not escape unscathed. As chief engineer he answered a subpoena to appear before the grand jury, where District Attorney Jerome attacked him for dereliction of duty for not solving the Park Avenue tunnel problem. One newspaper editorial sarcastically described Wilgus as "only thirty-six and not a graduate of a technical school."[8] Nothing wounded him more deeply than to have his professional competence challenged or, worse, to have someone question whether he had the basic competence to fill the senior engineer position with one of the country's largest railroads.

The following year in May, New York State and the city of New York passed legislation prohibiting the use of steam locomotives for passenger trains south of the Harlem River and in the Park Avenue tunnel. The legislation imposed a deadline of July 1, 1908. Not only could the Central no longer use steam power, but without a solution for the railroad's myriad problems at Grand Central, the New York Central would forfeit the cornerstone of its entire passenger system—access to midtown Manhattan at 42nd Street.

Wilgus knew that, above all, the New York Central had to keep its terminal at 42nd Street. The location held numerous advantages, not the least being that this was the only passenger terminal on Manhattan Island. The Central's chief rivals, including the Pennsylvania Railroad, all deposited their long-distance passengers on the New Jersey shore, where they boarded ferries to complete their journey to Manhattan. In addition, the 42nd Street depot also served as the hub of the railroad's suburban commuter service. By 1900 the number of commuters to

Grand Central each day almost matched the number of passengers riding the railroad's long-distance trains. For most commuters their journeys did not end at 42nd Street but farther downtown, with many working on Wall Street. For commuters traveling to New York's business and manufacturing districts farther downtown, 42nd Street brought them much closer to work than would a station in Mott Haven, which was north of the Harlem River and would require a long trip down the island on the elevated railroads or horse-drawn streetcars.

A Plan of Staggering Complexity

Immediately after he had arrived in New York, Wilgus had begun to study the complicated set of intertwined problems the Central faced with its outmoded steam-powered depot at 42nd Street. An open train yard at grade level filled the blocks to the north. Wilgus described the open train yard as a "veritable 'Chinese' wall to separate the city into two parts for fourteen blocks—nearly three-quarters of a mile—between 42nd Street and 56th Street."[9] On 5th Avenue, just two blocks to the west, lived the wealthiest New Yorkers, including the extended Vanderbilt family. On the other side of the "Chinese wall," to the east, tenements filled the blocks to the East River. On both 2nd and 3rd Avenues the city's elevated railroads ran north and south, providing mass transit up and down the island but creating horrible conditions for the immigrants forced to rent apartments along the avenues. Night and day the steam-powered elevated railways filled the air with smoke, steam, and screeching noise.

Wilgus began to envision a radical solution to the Central's interwoven problems at 42nd Street. Rather than a piecemeal plan to deal with the immediate crisis, he proposed a complex, integrated plan to replace the old depot and train yard and switch motive power from steam to electricity. Above the new terminal he proposed that the railroad construct a twelve-story office building to generate rental income. Even more boldly, Wilgus planned to take the entire open train yard and place it underground and build not just one train yard, but a two-story, underground train yard powered by electricity. Never modest, he described his plan, referring to himself in the third person, as a "Concept of an Entirely New Terminal" and added that "as he studied various proposals for retaining the old station in the new project, the writer was not satisfied that this was the ideal solution. . . . Why not (he questioned himself) tear down the old building and train shed and in their place, and in the yard to the north, create a double-level, under-surface terminal . . . made possible by the intended use of electric motive power."[10]

On December 22, 1902, Wilgus sent a letter to the railroad's president, William Newman, outlining his plans for the Grand Central project. In just three pages he presented a bold, imaginative plan for transforming a ten-block area in midtown Manhattan into the most complex railroad terminal in the world.[11] Wilgus followed with a letter in March 1903 laying out the details of his plan to transform midtown Manhattan and improve the Central's service throughout the metropolitan region.[12] The features included

- a two-story, electrified underground train yard with loop tracks from 42nd to 56th Streets between Lexington and Madison Avenues;
- a new terminal building on 42nd Street with central concourses on two levels—an upper level for long-distance trains and a lower level for suburban commuter trains;
- twelve stories in the new terminal building for rental space;
- ramps for passengers from the main concourse and to platforms for long-distance trains;
- ramps and elevators from the main concourse to the lower-level suburban concourse and then ramps from the lower-level concourse to platforms for suburban trains;
- north of the new terminal building, between 45th and 48th Streets, a separate baggage facility connected to the tracks below by elevators;
- a first-class hotel, connected to the new terminal on Madison Avenue between 43rd and 44th Streets;
- North of the terminal, over the underground train yard, provisions for future development of other revenue-producing buildings;
- an elevated roadway to carry Park Avenue from 42nd Street around the new terminal building and north over the underground train yard, restoring Park Avenue as a major north-south artery; and
- electrification of the tracks in Manhattan north of the underground train yard, in the Bronx, and into Westchester County to Croton-on-Hudson and North White Plains.

Wilgus also provided a cost estimate broken down into four components:

Underground train yard	$8,000,000
New terminal building	13,000,000
Electrification	10,000,000
Miscellaneous	4,000,000
"Roughly estimated total costs of improvements"	$35,000,000

For the New York Central Railroad, committing to a project of this magnitude represented an enormous undertaking. To put Wilgus's estimate into perspective, the total revenue of the Central for the 1901–2 fiscal year totaled $70,903,868.[13] After deducting expenses, the railroad realized net earnings of $28,916,402. From net earnings came interest payments on bonds and a 5 percent dividend, leaving the Central with a surplus of just over $2 million. Thus, Wilgus's initial cost estimate for the Grand Central Terminal project totaled almost exactly 50 percent of total revenue for one entire year.

While the railroad's passenger operations at 42nd Street stood at the heart of the company's passenger system, its main-line track system was much larger, totaling over 8,000 miles. The Central made much more money hauling freight than passengers. For example, freight operations in 1901–2 generated $40,659,778 in revenue, versus $23,807,085 from passenger business. Nevertheless, at the beginning of 1903 Wilgus was proposing spending $35 million on the Grand Central project, serving only the railroad's passenger business.

Wilgus was also proposing to President Newman and the New York Central Railroad one of the largest and most complicated construction projects in New York City's history. The multiple components of the plan would all have to fit together to solve the problems the railroad faced with its open train yard and aging passenger depot. In the background loomed the state-legislated mandate to convert from steam to electric power within five years. Wilgus's supreme confidence in his own abilities and in the application of sound engineering principles and rational planning would see the project through to a successful conclusion. Wilgus had no doubts that he would succeed, viewing his plans in heroic terms:

> I had no hesitation in now proceeding. . . . The bars were down to the permanent solving of the Park Avenue–Grand Central problem . . . [and] marked the opening of a remarkable opportunity for the accomplishment of a public good with considerations of private gain in behalf of the corporation involved. In the classic words of the founders of the civil engineering profession, as such, one of its members in this case [Wilgus] was now free to direct one of the great sources of power in nature here evoked for the use and convenience of Man.[14]

Air Rights

A decision to proceed with a project of this magnitude did not rest with President Newman. The railroad's board of directors would have to make the final decision. In 1903 the board included two grandsons of Cornelius Vanderbilt, William K. and Cornelius III; William Rockefeller, brother of John D. Rockefeller; and J. P. Morgan—all of whom wanted to protect the financial resources of the railroad. J. P. Morgan above all demanded careful planning and fiscal soundness before the Central undertook any major capital improvements. Morgan served on the boards of numerous railroads, and his company, J. P. Morgan & Co. provided all of the Central's financing. Morgan had joined the boards of many railroads following the financial panic of 1893, when numerous railroads declared bankruptcy, destroying their investors. Morgan would not support such an extensive project unless Wilgus convinced him and the rest of the board that the commitment of such an enormous amount of money represented a good investment. But Wilgus had already conceived of a way for the railroad to carry out the entire Grand Central plan on a self-financing basis.

Wilgus realized that "air rights" provided the resources to finance the entire Grand Central project: "The keynote in this plan was the utilization of air rights that hitherto were unenjoyable with steam locomotives requiring open air or great vaulting spaces, for the dissipation of their products of combustion. Thus from the air would be taken wealth with which to finance obligatory vast changes otherwise nonproductive. Obviously it was the thing to do."[15] A magnificent new terminal building, the two-story underground train yard, and electrification could all be paid for by taking "wealth from the air."

In part this would be done by constructing a new twelve-story office building above the main terminal concourse. By 1903 the incessant growth of Manhattan surrounded the depot and train yard on 42nd Street. Office space in Midtown was becoming more desirable. Wilgus pointed out that the Central itself rented office space just across the street for $1.31 a square foot. The office building proposed in Wilgus's plan would contain 2.3 million square feet of rental space. If one assumed, conservatively, that this space could be rented for $1.00 per square foot, the new building would provide $2.3 million in rental income per year. More importantly, north of the new Grand Central Terminal were multiple city blocks now occupied by the open train yard. By switching from steam to electric power and constructing an underground train yard, the New York Central could construct apartments and office buildings above the train yard to generate further revenues.

Taking wealth from the air was an inspired idea for a railroad facing huge expenses at 42nd Street. In a stroke of genius Wilgus not only envisioned a complex, multi-part plan for the new Grand Central, but also outlined a plan to finance the entire project.

An article in 1912 in the *Railway Age Gazette,* a year before the Grand Central terminal finally opened, pointed out the advantages to be gained from utilizing air rights: "The New York Central & Hudson River . . . have been making plans to utilize the space above their terminal tracks in such a way as to make the income from this space pay the entire interest charges on the cost of building the terminal yards and even the station itself." The article discussed "the unlimited demand for office space in this neighborhood" and the "unusual transit facilities which are concentrated at this part of New York City."[16] Numerous street railways, the city's first subway line, and an extension of the 3rd Avenue Elevated Railroad all converged in front of Grand Central. No other location in the city offered the same transportation nexus.

Electricity

All of the proposed changes at 42nd Street depended upon the successful switch from steam power to electricity, allowing the New York Central to construct a two-story underground train yard, to replace the open train yard that created a barrier for ten blocks, and to eliminate the hazard of steam and smoke in the Park Avenue tunnel as well as in the planned two-story underground train yard.

Electricity created a revolution in American society, changing every aspect of life for the part of population with access to electric power.[17] In 1880, Thomas Edison, already world-famous, formed the Edison Electric Illuminating Company of New York and received a franchise to provide incandescent lighting for customers in lower Manhattan. Edison built a power plant on Pearl Street in lower Manhattan and after hectic efforts transmitted the first electric power for indoor lighting on September 4, 1882. Prominent among Edison's customers was the firm of J. P. Morgan & Co. on Wall Street. Building electric power plants and transmission lines and providing reliable service proved to be very expensive. For the next three decades electricity remained a luxury only a few businesses and wealthy individuals could afford.

Using electricity for transportation necessitated a great deal of innovation. Electricity first had to be generated at a central power plant and then distributed in a safe manner over sizable distances. To power street cars or subway cars required electric motors capable of converting the electricity to mechanical power providing traction to move the cars over the tracks. Edison, Frank Julian Sprague, Bion Arnold, and the other pioneering electrical engineers worked furiously to develop electric "traction" to convert mass transportation from horse and steam power to electricity.

Frank Sprague, another genius of the age, was born in 1857 in Milford, Connecticut, and graduated from the U.S. Naval Academy in 1878 with an engineering degree. Serving briefly as an active officer, Sprague left the Navy in 1883 and went to work for Thomas Edison in his famous laboratory in Menlo Park, New Jersey. After just a few years with Edison, Sprague formed his own company to build electric-powered street railways. Initially, Sprague tried to interest the Manhattan Elevated Railroad in substituting electric power for steam. The els in New York carried huge numbers of passengers each day up and down the major avenues, but the company failed to see the advantage in switching from steam to electric power, still an unproven technology.

Sprague continued to experiment with electric power and finally secured a major contract in 1887 to construct the nation's first electric-powered street railway in Richmond, Virginia. Sprague described the challenges he faced: the ninety-day contract called for "twelve miles of track . . . ; the construction of a complete steam and electric central-station plant of three hundred and seventy-five horse-power capacity; and the furnishing of forty cars with eighty motors." "This was," he noted, "as many motors [as were] in use on all the cars throughout the rest of the world."[18] Sprague would be paid $110,000, but only "if satisfactory." He decided to transmit the electricity by overhead wires and equipped each car with a contact pole on the roof to conduct the current from the overhead wire to the undermounted electric motors. With supreme confidence Sprague succeeded, and in January 1888 the Richmond electric street railway began service. In the space of a few years, 110 other cities across the country installed electric-powered street railways.

While Sprague's success in Richmond sparked an electric street railway frenzy across the country, the electrification of the New York Central's passenger operations in New York represented a challenge of an entirely different magnitude. The Richmond street cars, fully loaded, weighed about two tons; a 20-horsepower electric engine provided all the needed power. By comparison, the heavy long-distance passenger trains using Grand Central weighed hundreds of tons and required electric engines capable of generating thousands of horsepower. A single electric power plant in Richmond with a modest output provided enough electricity to power 40 street cars, but 500 trains arrived and departed from Grand Central each day. If Wilgus and the Central switched from steam to electric power, the engineering required more than simply scaling up an electric street railway. Completely new equipment capable of moving hundreds of tons of passenger cars and a massive generation and transmission system would have to be carefully planned, tested, and then constructed.

The Electric Traction Commission

To make it possible for the New York Central to switch to electric power, Wilgus proposed to President Newman the establishment of a commission, the "Electric Traction Commission," to plan for and oversee the switch. A model of rational planning, the commission included Wilgus, two other senior New York Central engineers, and three outside consultants. Wilgus asked Frank Sprague, Bion Arnold, and George Gibbs, among the most renowned electrical engineers in the country, to serve as consultants; all accepted the offer.

As a first step, Wilgus asked Bion Arnold to conduct a preliminary study of the overall feasibility of converting from steam to electric power. Initially, Wilgus asked Arnold to focus on switching to electricity only on Manhattan Island between Grand Central and Mott Haven, just across the Harlem River in the Bronx, as the state mandate required. Arnold submitted a report to Wilgus and in his cover letter concluded: "It is feasible and practicable to equip and successfully operate, electrically, the section of the New York Central Railroad extending between Mott Haven Junction and Grand Central Station." Arnold estimated the cost at $2.5 million.[19]

Arnold's report provides a model of the rational professional planning which characterized the Grand Central electrification. Wanting to leave as little as possible to chance, Wilgus had asked Arnold to carefully study all aspects of the proposed switch to electric power. Several engineers, among them Arnold and Wilgus, first collected relevant data, performed the necessary calculations based on sound scientific principles and then developed recommendations based on the analysis. Arnold's report also included a detailed cost analysis and laid out all of the major issues the railroad needed to address in order to switch to electric power.

To estimate the amount of electricity needed for operating the number of trains currently traveling to and from Grand Central, Arnold employed a "dynamometer" car to calculate the horsepower needed to haul the various classes of passenger trains. The University of Illinois and the Illinois Central Railway jointly owned the dynamometer car, and Arnold hired Prof. E. C. Schmidt of the university's Railway Engineering Department to run the tests.

Arnold and Schmidt coupled the dynamometer car between the steam engine and the first passenger cars on a sample of the Central's 500 scheduled passenger trains operating each day at Grand Central. The dynamometer measured the drawbar pull for different size trains. Arnold converted the pull to horsepower and then to the number of electric kilowatts needed to supply the needed horsepower. He estimated an average of 1,800 kilowatts per hour to power all trains and a total consumption of 15,768,000 kilowatt hours for 205,285,710 ton-miles of service.[20] Arnold's careful analysis provided Wilgus and the Electric Traction Commission with the data to plan for the needed electric generation capacity.

At the first meeting of the Electric Traction Commission on December 17,

1903, Wilgus laid out an ambitious agenda. Among the pressing issues to be decided he included the following:

- extent of electrification (Manhattan Island versus an "electric zone" through the Bronx and into Westchester County);
- alternating versus direct current;
- number of electric power plants;
- distribution of electricity by third rail versus overhead power lines;
- design and specifications for new electric engines.[21]

Over the next year, the Electric Traction Commission met 103 times and carefully evaluated all relevant data, engaged in open debate, and then made a series of crucial design recommendations to the senior management of the railroad.

Crucial questions about the design of the electric system required ingenuity and boldness. No existing railroad electrification anywhere in the world approached the scale of the Central's project or provided a model to duplicate. Much of the equipment, especially the new electric engines for long-distance trains, required design, testing, and manufacture from scratch.

The legal mandate to switch from steam to electric power applied only to Manhattan Island, and initially the Central planned to switch from steam to electric engines at Mott Haven, just over the Harlem River in the Bronx, nine miles north of Grand Central. Arnold's preliminary study analyzed only electrification between Mott Haven and 42nd Street. From his very first plans submitted to Newman, Wilgus envisioned a much more extensive "electric zone" extending well into Westchester County, on the west; to Croton-on-Hudson, thirty-three miles from Grand Central; and on to White Plains, twenty-three miles from Grand Central.

Initially, he faced strong opposition to an expanded electric zone from both Bion Arnold and Frank Sprague. Arnold's study included a list of existing electrified rail systems, all of which ran for comparatively short distances and did not involve heavy long-distance passenger trains (table 2.1). Arnold worried that expanding the electric zone required untested transmission technology and dramatically increased the cost of electrification.

Many of the electrified railways listed by Arnold—the Albany & Hudson, the Providence & Fall River, and the Nantasket Beach, for example—consisted of modest one-track streetcar railways running a few cars each day. In many places north of Mott Haven, on both the Hudson and Harlem Divisions, two tracks paralleled each other, and in some areas four tracks ran in parallel—all to be electrified.[22] An article in the *Railroad Gazette*, analyzing the electrification, reported the total miles of track electrified at 292, including 224 miles of main line tracks and 68 miles of yard tracks.[23] The Grand Central electrification involved building the largest electric-powered rail system in the world. All of the electrified rail lines listed by Arnold totaled just 212 miles of main line tracks. The New York Central planned to construct an electric zone in the New York metropolitan region larger than all of the electric-powered railways in the world combined.

Wilgus characterized the debates of the Electric Traction Commission as

Table 2.1. Electrified rail lines in 1902

Railway	Year	Distance (miles)	Type of current	Means of electrical distribution
United States				
Providence & Fall River, RI and MA	n.r.	18	Direct	Overhead wires
Baltimore & Ohio, Baltimore, MD	1886	3	Direct	Overhead wires
Buffalo, Tonnawanda, NY	1889	13.5	Direct	Overhead wires
Lake Street Elevated RR, Chicago	1895	11	Direct	Third rail (side)
Metropolitan Elevated RR, Chicago	1895	18	Direct	Third rail (side)
Nantasket Beach, MA	1897	10.5	Direct	Overhead wires
Brooklyn Elevated RR, Brooklyn	1898	n.r.	Direct	Third rail (side)
South Side Rapid Transit, Chicago	1898	8	Direct	Third rail (side)
Albany & Hudson, NY	1900	37	Direct	Third rail (side)
Northwestern Elevated RR, Chicago	1900	8	Direct	Third rail (side)
Boston Elevated RR, Boston	1902	8	Direct	Third rail (side)
Britain and Europe				
Burgdorf–Thun RR, Switzerland	1889	25	Alternating	Overhead wires
Central London Underground, London	1899	5.5	Direct	Third rail (center)
Stanstadt–Engleberg RR, Switzerland	1899	25	Alternating	Overhead wires
Gare de Austerlitz–Gare d'Orsay, Paris	1900	2.5	Direct	Third rail (side)
Jung-Frau Railway, Switzerland	1900	7	Alternating	Overhead wires
Metropolitan Railway, Paris	1900	7	Direct	Third rail (side)
Western Railway of France, France	1900	12	Direct	Third rail (side)

SOURCE: Bion Arnold, "Report of Bion Arnold to William Wilgus upon the Proposed Electric Equipment of the Hudson Division of the New York Central Railroad," 3–7, Bion T. Arnold Papers, Manuscripts Division, New York Public Library, box 6.
NOTE: n.r. = not reported

"prolonged arguments, often heated," but the commission finally approved the extended electric zone on November 3, 1903.[24] Part of Wilgus's argument for an expanded zone involved the railroad's commuter business, which was growing substantially each year. Providing commuter service with steam-powered trains created operating difficulties. Steam engines accelerated slowly and, because of short distances between suburban stations, could not provide fast and efficient service. With the switch to electric service for the commuter trains, Wilgus projected that the number of daily commuters would increase, generating significant revenue to offset the increased fixed costs of constructing a much larger electric zone. His views prevailed: "The Commission unanimously decided that the plan as first promulgated [by Wilgus] should stand unchanged—namely that both multiple-unit suburban trains and express [long-distance] trains hauled by electric engines should be operated for the full distance between the Grand Central Terminal and the northerly terminals at Croton-on-Hudson and North White Plains."[25]

The "multiple-unit" suburban trains referred to were the electric-powered suburban passenger railcars planned for the suburban commuter service. Plans called for each passenger car to have an electric traction motor and be self-propelled. The railroad could assemble trains made up of a few cars or multiple cars depending on the number of passengers. During rush hours, ten or twelve cars would be coupled together and filled with morning commuters heading for work in New York or evening commuters returning to their suburban homes. Non-rush-hour trains would require only three or four passenger cars.

Of course, coupling multiple self-propelled suburban trains presented another major challenge. The electric motors in each of the coupled suburban passenger cars would need to be simultaneously controlled to ensure smooth and safe train operations. In 1897 Frank Sprague patented a system to operate multiple electric traction motors in tandem. Sprague, a pioneer in electric elevator controls, realized that his master control system for elevators could be employed to revolutionize railway electrification:

> Soon after taking up the development of electric elevators, I made the distant control of the main motor-controller from a master switch a *sine qua non* for all important work. . . . Pondering over the railway train problem one day, the thought suddenly flashed upon me, Why not apply the same principle to train operations. That is, make a train unit by the combination of a number of individual cars, each complete in all respects, and provide for operating them all simultaneously from any master switch on any car. The idea, sketched on a scrap of paper, marked the complete birth of this new method, then named and now nearly everywhere known as the "multiple unit method." . . . I saw the opening of a new epoch in electric railway operation. Here was a way to give a train of any length all the characteristics of a single car.[26]

New York City's subways, subways around the world, and electric commuter railroads all adopted this "multiple-unit" control system. Sprague's invention created a new era in railroad electrification, especially for suburban commuter service.

Deciding whether to use direct or alternating current also created challenges for the Electric Traction Commission and involved the "battle of the currents." Pioneering efforts in electrification, led by Thomas Edison, utilized direct current. To create a power system, a generating source produced direct current at a certain voltage—100 volts, for example—and then distributed the electricity through wires to an incandescent light or an electric motor consuming 100 volts of electricity to provide light or raise and lower an elevator car. Edison's first power plant on Pearl Street; the city's first subway line, the IRT (Interborough Rapid Transit Company); and Sprague's electric street railway in Richmond—all used direct current. But direct current has a serious limitation: it cannot easily be distributed over long distances without losing power owing to the resistance in the transmission wires.

Alternating current, current which switches direction back and forth, can be transmitted over long distances by using a transformer to increase the voltage at the generator and then to decrease the voltage where the power is used. Advocates of railroad electrification envisioned all of the major trunk line railroads in the United States switching to electric power, which would require the transmission of electricity over long distances, favoring alternating current. Bion Arnold recognized the advantages of alternating current: "I believe that the alternating current railway motor will prove to be the most efficient, all things considered, for long distance railway work." On the other hand, he also noted that alternating current "has not yet demonstrated its ability to start under load as efficiently or to accelerate a train as rapidly as a direct current motor."[27] For the suburban commuter service in the electric zone, the railroad needed electric cars to accelerate rapidly, stop at the next station a few miles down the line, and then accelerate again smoothly and quickly. Arnold recommended direct current, and Sprague supported the recommendation.

The battle of the currents involved not just electrical engineers; powerful corporate interests also took sides. The Westinghouse Corporation, led by the renowned inventor and entrepreneur George Westinghouse, advocated alternating current for electric traction. Westinghouse manufactured alternating current equipment. General Electric, which had acquired most of Edison's electric businesses in 1889, supported direct current.

As soon as the New York Central announced its decision to use direct current for the Grand Central electrification and had awarded the contract to build the new electric engines to General Electric Company, the battle of the currents erupted in public. George Westinghouse wrote an "open letter" to President Newman of the Central in the *Railroad Gazette*, arguing that the railroad made a serious mistake in choosing direct current. He asserted that a direct current system would cost much more per mile than alternating current. Never hesitant, Westinghouse demanded the Central use alternating current equipment, presumably, of course, to be supplied by the Westinghouse Electric Corporation. Not satisfied with questioning the judgment of the president of the New York Central, Westinghouse also openly attacked Frank Sprague, accusing him of a conflict of interest. General Electric had acquired the rights to Sprague's multiple-unit control system when it purchased the Sprague General Electric Company in 1901. Westinghouse charged that Sprague would benefit secretly if General Electric became the prime contractor for the Grand Central electrification and that this influenced his support on the commission for using direct current equipment to be supplied by General Electric.[28]

Sprague responded immediately with his own "open letter" in the *Gazette*, defending his professional reputation in no uncertain terms: "I would brook no interference by individual or corporation with my professional opinion. . . . My engineering convictions and conclusions are my own. They are dictated by no man or corporation."[29] Sprague, Arnold, and Wilgus all saw themselves, first and foremost, as professional engineers. Their world consisted of hard data, sound scientific principles, careful analysis, and rational decision making—all for the public good and not just for narrow corporate or personal profit. Challenge their professional integrity and they fought back. Sprague reminded Westinghouse in his open letter that his corporate empire and personal wealth rested on the inventiveness of engineers, including Sprague himself.

Next, the Electric Traction Commission unanimously recommended distributing direct current in the electric zone by a third-rail electrification system. A third-rail system added a rail next to all of the tracks to be electrified. Electric current—in this case, direct current—flowed through the third rail. A "shoe" extending out from the electric engine made contact with the third rail, and electricity flowed through the shoe to the electric motor. As Arnold's study documented, most of the existing electric railways distributed current by a third rail; these included New York City's first subway, nearing completion in 1906.

Not to let matters rest, Westinghouse also objected to the choice of a third-rail system and strongly urged the Central to use overhead wires for electric distribution, even if it insisted on proceeding with direct current. Westinghouse and other critics worried about the threat to the safety of railroad workers and passengers posed by a third rail. The possibility existed for railroad workers, going about their regular work, to accidently come into contact with the third rail and be electrocuted. In the electric zone, the third rail would carry 660 volts, a potentially lethal amount of electricity.

Wilgus and Sprague answered the critics with a clever invention for which they submitted a patent application in 1905: a wooden cover over three sides of the third rail, the top and the two sides. Only the bottom of the third rail remained exposed. A spring-loaded shoe from the electric engine rode under the third rail, maintaining contact with the rail and transferring electric power to the engine. To accidentally come into contact with the electricity in the third rail, a railroad worker needed to touch the underside of the rail below the protective covering. A simple technology, the protective covering designed by Wilgus and Sprague countered the safety concerns raised by Westinghouse. To avoid any conflict of interest, both Wilgus and Sprague waived any payments from the New York Central for the use of their third rail protective covering in the electric zone.

Another important question before the Electric Traction Commission involved planning for electric power generation. While electricity had arrived in New York City in 1891, the electric industry remained fragmented, with a number of small companies competing to provide electric service. None of the electric companies had the capacity to supply the amount of electricity needed for the Grand Central electrification. Consolidated Edison, which eventually took over both the electric and gas industry in New York City, lay many years in the future.

Wilgus and the electrical engineers recommended building a failure-proof generation system with a great deal of built-in redundancy. With thousands of passengers riding the trains at any one time, a failure of the electric system would create chaos. The commission recommended building two separate power plants, one on the Hudson River in Yonkers and the second at Port Morris on the East River in the Bronx. Both power plants would be supplied with coal from either river barges or freight cars. Each of the "cross-connected power-stations," Wilgus wrote, would have "sufficient capacity . . . to carry the entire demand of the service at rush hours, should the other fail."[30] To further protect service, the commission also recommended building a number of storage battery buildings at eight substations, "probably the largest railway storage battery equipment in the world," reported *Railway Age*, "large enough to operate the entire train service under normal conditions for a period of one hour."[31] In the event that the power generation system failed, the batteries would take over and allow all passenger trains to safely reach the nearest station.

The commission solicited bids for electric generators from the two leading manufacturers, and again Westinghouse and General Electric faced off. Wilgus dryly remarked on the bidding that General Electric "with its lower price won the competition." Once again, George Westinghouse did not sit idly by and accept the careful, measured judgment of the Electric Traction Commission. A few days after the announcement of the awarding of the contact to General Electric, Wilgus received a surprise visit: "I received a call from Mr. Westinghouse, who on seating himself in my private office looked me in the eye and said, 'Young man, I am fond of you, and so it is with regret that I am moved to tell you that in favoring the [General Electric] turbo-generator you have lost your reputation.'"[32] Wilgus did not back down: "I had full confidence in the strength of my convictions that it was with difficulty I restrained a smile." Even with one of the giants of the age sitting across from him, he could not be bullied. Westinghouse did not rest and proceeded to visit the other senior officers of the railroad and voice his strong objections to the decisions of the commission.

One last crucial task for the Electric Traction Commission involved the design and specifications for the new electric engines to be used to haul long-distance trains. At its May 19, 1903, meeting, the commission carefully reviewed the final specifications for the electric engines, which called for the new engines to develop 2,500 horsepower and accelerate quickly while pulling a heavy load of passenger cars.[33] In addition, the electric engines had to be capable of a round trip of thirty-four miles in under one hour, pulling a 550-ton train.

Both General Electric and Westinghouse regarded the contract for the new electric engines as of great importance. They believed that the Grand Central electrification was the first wave of railroad electrification across the United States and would create a huge, lucrative market for electric railway engines. The company selected to provide the engines for the New York Central's electrification would assume a leading position. After carefully reviewing all the bids, the commission, in November 1903, recommended to the board of directors that the contract be awarded to the General Electric Company. The New York Central drove a hard bargain, demanding that General Electric agree to a cost of just over $30,000 for each of thirty engines in the initial order. In addition, General Electric had to complete the first engine in less than twelve months and subject the engine to a rigorous testing program on a special four-mile electric track set up outside of Schenectady. General Electric, delighted to have won the contract, agreed to all of the conditions imposed by the railroad and immediately set to work.

At the November 1, 1904, meeting of the commission, Wilgus proudly reported on the first successful run of the new electric engine a few days previously, on October 27. With eight passenger cars attached, the engine easily accelerated to 55 miles an hour. The engine also accelerated from a dead stop to a speed of 30 miles an hour in one minute, meeting a crucial specification imposed by the Central's contract. For the next three months, General Electric continued the testing program, operating the new engine eight hours a day on its test tracks at an average speed of 50 miles an hour, eventually running the engine a total of 45,000 miles.

The initial success of the new electric engine generated a great deal of attention, both in the major newspapers and in the technical press. A detailed article in the *Railroad Gazette* hailed the test as signifying a "new era in the development of transportation facilities in this country. . . . New York Central's electrification scheme is the first radical change on the part of an existing steam road to electric operation for comparatively long distances."[34] The publication reported that the 95-ton engine averaged a horsepower rating of 2,000 with a maximum of 3,000 horsepower, making it the most powerful electric engine in the world. The *Gazette* article included charts from two speed runs illustrating the dramatic capability of the electric engine. From a dead stop, the train, pulling four cars weighing 170 tons, accelerated to 60 miles an hour in just 160 seconds. With eight cars in tow weighing 336 tons, the engine accelerated to 60 miles an hour in 200 seconds, a little over 3 minutes.[35]

With Wilgus, the Electric Traction Commission, and the senior executives of the New York Central present on November 12, General Electric staged a dramatic race pitting the new electric engine against one of the New York Central's Atlantic-type steam engines running on a parallel track. Each engine pulled three passenger cars. From a dead stop, the electric engine pulled away from the steam engine and led by more than half a mile at the end of the four-mile test track. A headline the next day in the *New York Herald* read, "Electric Engine Beats All Rivals."[36] The normally calm, rational William Wilgus was overjoyed. Not only had General Electric completed its contract for the new engine on time, the engine met all design specifications set out by the Electric Traction Commission; Wilgus characterized the work of the company as "marvelous."

New York Central's first electric-powered train arrived in Grand Central with great fanfare and with Wilgus at the controls on September 30, 1906. Under Wilgus's direction and with the careful oversight of the Electric Traction Commission, the railroad had completed its revolutionary electric zone well before the July 1908 deadline imposed by New York State and the city. The *World* hailed the success of the first electric engine to Grand Central with a headline: "First Electric Train Enters New York City, Thousands of Spectators Cheer."[37] The first scheduled electric suburban train left Grand Central on December 11, 1906, and the first long-haul train followed in February of 1907. By July 1907 all scheduled service to Grand Central had been switched to electric power. Wilgus, in an understated and matter-of-fact manner, summed up the company's success: "The New York Central by July 1, 1908 well kept its promise to abolish the long-standing smoke, gas, and cinder nuisance in Park Avenue [tunnel] and at its

First electric engine to Grand Central Terminal, September 30, 1906, with Wilgus at the controls. The old train shed and Grand Central Depot are in the background.

terminal yard and station."[38] Wilgus's self-confidence and vision had served as the driving force behind one of the most important technological achievements in New York City history.

The work of the Electric Traction Commission represented the triumph of systematic planning and organization, the careful, deliberative application of engineering to the implementation of a most complex project. At each stage of the planning process the commission gathered relevant data, conducted necessary testing, and weighed the alternative choices. David Nye, a renowned historian of technology, emphasizes the role choice played in the process of electrification: "Such technological developments are too often understood as irresistible, when in fact people shape the form of the electrical system."[39] The careful planning of Wilgus and the commission decisively shaped the final configuration of Grand Central electrification.

Construction

At the same time that the New York Central was making crucial decisions about electrification and the electric zone, the railroad initiated four interconnected construction projects. First, the O'Rourke Construction Company started excavation for the underground train yard. Demolition of Commodore Vanderbilt's aging depot followed, and then, in the fall of 1911, construction of the new terminal building on 42nd Street commenced. When the new underground train yard was completed, construction of the "air-rights" buildings to the north along Park Avenue began and continued for the next two decades.

Just as with the revolutionary electric zone, the scale of the construction

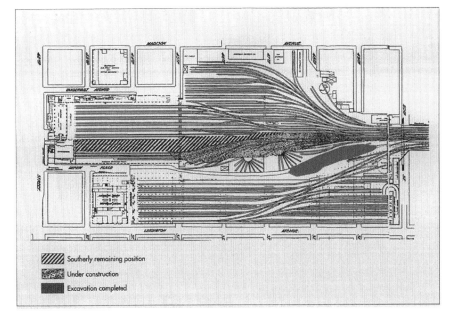

Wilgus's map for the first stage of construction of Grand Central Terminal—the planned excavation of the open train yard, 1903.

- Southerly remaining position
- Under construction
- Excavation completed

would be unprecedented. To make matters even more complicated, the new construction could not interfere with the operation of the busiest railroad terminal in the country. All scheduled passenger service, long-haul and suburban, an average of 500 trains a day, had to be maintained without interruption. Therefore, Wilgus devised an ingenious plan to carry out the huge construction project as a series of "bites" proceeding from Lexington Avenue on the east to Madison Avenue on the west. Each bite involved

1. demolition of any surface-level structures
2. excavation for the two-story underground train yard
3. construction of the lower- and upper-level underground train yard and platforms
4. electrification of the new tracks
5. restoration of train service using the new section of the underground train yard

Wilgus planned a total of eleven "bites" with the first one along Lexington Avenue, where demolition and excavation got under way in the summer of 1903. As work ended in each bite, control of the completed section passed to the operating division, train operations commenced, and the construction crews moved on to the next bite to the west. Wilgus created a detailed time line for each of the twelve bites with work in bite 1 to be completed by December 1905 and the last bite to the west by January 1908.

To manage the complicated work, Wilgus set up a construction committee to parallel the work of the Electric Traction Commission. The committee's members consisted of the outside architects, the heads of the operating divisions of the railroad, and representatives of the excavation and construction companies hired to do the actual work. The Construction Committee met frequently over the course of the construction effort and carefully monitored progress. As major difficulties arose—inevitable with a project as large as Grand Central—the committee analyzed the sources of the problems and decided on appropriate courses of action.

Creating a two-story underground train yard required the excavation of over three million cubic yards of earth and rock, to a depth of 90 feet below the existing grade level. Since much of Manhattan Island consists of hard rock, blasting was necessary to break up the rock to allow for removal. In August 1903, after careful evaluation of bids, the railroad awarded the excavation contract to the O'Rourke Construction Company of New York at a cost of $8,550,000 for all work south of 57th Street.

From the very first days of the excavation, problems arose with the work of the O'Rourke Company, and the minutes of the Construction Committee chronicle the frustration of Wilgus and the other Central executives with the pace of the excavation. By the spring of 1907, Wilgus and the Construction Committee had lost all patience with the inability of O'Rourke to meet deadlines. Wilgus recommended to the board of directors canceling the contract with O'Rourke and paying the company for all work completed to date.

In place of O'Rourke, Wilgus and the railroad decided to transfer the excavation of the underground train yard to the Central's own work forces, and work on the train yard continued. Like all major trunk line railroads, the Central employed a large construction work force to maintain tracks and rail yards. Railroad engineers regularly managed large construction projects as part of the essential work of running the railroad.

Wilgus listed seventy-three major contracts for the train yard and the construction of the new terminal building awarded between 1903 and 1907. With a project as complicated as the one unfolding in midtown Manhattan, plans seemed to be in constant flux, and a massive construction effort was involved. The train yard alone required the excavation of over 3 million cubic yards of rock and earth, construction of 260,000 cubic yards of masonry, and the use of over 100,000 tons of steel. In addition, construction crews rebuilt or relocated twenty-seven miles of pipes and sewers, built twenty-seven miles of tracks, and created the most complicated switching and signal system in the world, with 762 switch levers.[40] Passenger platforms on both the upper long-distance and lower suburban levels totaled six miles in length.

Inevitably with a project of this magnitude, costs increased. Until he left the railroad in 1907, each year Wilgus presented a detailed report to President Newman with the updated cost estimates. When he first outlined the scale of the project, he had projected total spending at $34,360,000 for eight separate components:

1. excavation, enlargement, and construction of the two-story underground train yard

2. construction of the Grand Central Terminal building
3. electrification: underground train yard, Park Avenue tunnel, electric zone to Croton-on-Hudson and North White Plains
4. improvement of Port Morris branch line in the Bronx
5. straightening of Marble Hill tracks in the Bronx, connecting Hudson Division to Harlem Division
6. elimination of grade crossings and depression of tracks in Highbridge, Morris Heights, Fordham Heights, and the Bronx
7. Hudson Division: addition of four-tracking to Croton-on-Hudson
8. Harlem Division: addition of four-tracking to White Plains

From Wilgus's first estimate in 1903 to his last in 1906, overall costs increased by over 100 percent, to a breathtaking total of $71,825,000. While Wilgus included detailed information to explain the rise in costs, President Newman and the board of the railroad faced a dilemma. Once the project commenced and the company signed agreements with the city and state to eliminate the use of steam motive power and convert to electric operations, the Central had no choice but to press forward. Since retained earnings could not cover even a fraction of the escalating costs, only one alternative remained: to borrow ever-increasing amounts of capital and complete the project.

The Woodlawn Wreck

At the height of his career, with the massive Grand Central project well under way, William Wilgus abruptly resigned from the New York Central Railroad in July of 1907. Since joining the Central's subsidiary, the Rome, Watertown & Ogdensburg Railroad in 1893, he had enjoyed nothing but success. In just fourteen years he had risen to the pinnacle of railroad engineering in the United States and had initiated the most complex railroad construction project in any American city. The Central rewarded his efforts in January 1907 by raising his salary to $40,000, placing him, at the age of 40, among the highest-paid railroad executives in the country.

Wilgus, a proud man, perhaps somewhat arrogant, valued his professional reputation highly and never stood for any challenge to his competence. In the aftermath of the Park Avenue tunnel wreck in 1903, he was outraged when the press questioned his competence to serve as the chief engineer of the Central and made reference to his lack of a college engineering degree. On the night of February 15, 1907, another New York Central train wreck, at Woodlawn in the Bronx, again challenged his reputation.

One of the new electric-powered trains, an express on the Harlem Division, departed from Grand Central for White Plains at 6:15 p.m. At 205th Street in the Bronx, rounding a curve, the train derailed. Rescuers arrived quickly on the scene and found twenty-five dead and close to one hundred passengers injured, some horribly, in the wreckage scattered along the tracks for over a mile.[41] Some newspapers were full of inaccurate reports: that many of the dead had been electrocuted and that other victims had burned to death, trapped in the wooden passenger cars set on fire by contact with the electrified third rail.

An uproar ensued. Assistant District Attorney Smyth called for the indictment of the Central executives for manslaughter and vowed to investigate the accident vigorously. The good will and laudatory press the railroad had received in the switch from steam to electric power in the fall of 1906 evaporated.

For William Wilgus the Woodlawn wreck represented a watershed in his career with the Central and a turning point in his life. He learned of the wreck while on a much-deserved vacation in California. After more than four years of nonstop efforts directing the giant Grand Central project, including direct oversight of the crucial electrification, he requested a month-long leave of absence. When the wreck occurred, President Newman telegraphed Wilgus and demanded that he immediately return to New York. Wilgus believed his entire professional career rested on his ability "to defend the electric installation for which the primary responsibility was mine."[42]

As devastating as the criticism and condemnation in the popular press proved to be, the railroad also faced a threat to the entire electrification project, the cornerstone of the Grand Central project. An editorial in the *New York Journal* placed the blame for the wreck on the design of the new electric engines—the design formulated by Wilgus and the Electric Traction Commission. Critics claimed the design concentrated the entire weight of the new electric engine on the drive wheels and caused the tracks in the Bronx to spread apart, derailing the train. If the design of the electric engines caused the wreck, the New York State legislature might pass legislation outlawing the use of the new engines and suspending the Central's entire electrification effort. In that case the railroad faced financial ruin.

As Wilgus hurried back to New York, senior officers of the Central rushed to blame each other for the disaster. Testifying before the State Railroad Commission, President Newman, Vice President Brown, and the head of the Operating Division, Alfred H. Smith, placed the blame on Wilgus and the Electric Traction Commission. Following this testimony, District Attorney Smyth issued a statement to the press calling for Wilgus to be indicted for manslaughter.

With the help of his staff Wilgus assembled materials to defend himself and the work of the Electric Traction Commission. During this process he learned of a neglected report of a track defect at the exact location of the Woodlawn wreck which shifted responsibility to the Operating Division and its maintenance-of-way, the subdivision responsible for the proper upkeep of the railroad's track. With this evidence, along with "other facts gleaned from members of my organization and from the files," Wilgus wrote in his autobiography, "I was in a position to convince the coroner that the electric installation *per se* was guiltless."[43]

After his testimony Wilgus thought that both President Newman and Senior Vice President W. C. Brown of the Central now believed the fault lay with the maintenance-of-way and not with the design of the electric engines or any of the work of the Electric Traction Commission or his own efforts. For the moment he felt vindicated.

To his complete surprise, in May, Vice President Brown remarked to Wilgus that he and other senior officers now believed the Woodlawn wreck to be the fault of the electrification, specifically the design of the new electric engine. His technical and engineering competence challenged, Wilgus responded in his typical analytical fashion. He immediately set about assembling a "Woodlawn wreck" report, with detailed information from his files and from the exhaustive testing program conducted with the new engine in Schenectady, New York.

Part of the report discussed the attention given during testing to a "nosing" problem (the term railroads used to refer to an engine's weight spreading apart the tracks). A very serious problem, nosing could lead to a derailment like the one that occurred in Woodlawn. General Electric and the American Locomotive Company engineers had documented such a problem and had modified the wheel design. After the design change, both companies certified "that in their judgment no detrimental nosing would occur in the New York District."[44]

Whether Wilgus realized it or not, his Woodlawn wreck report represented a time bomb for the New York Central Railroad and its senior officers, Wilgus included. If they knew beforehand of the potential for nosing, did they exercise due diligence before putting the new engines in daily service in the electric zone? Wilgus described in detail the events which followed: a corporate cover-up at the highest level of the railroad. After he submitted his report with its supporting materials, he received a visit "from General Counsel Place, a long-time friend and associate, with a message from Mr. Newman and Mr. Brown . . . that were this report to appear in public 'some one would go to jail.' In response to his plea that I should consent to its destruction in the company's interest, I did so with the proviso that all concerned should recognize that my skirts had been cleared, and that Mr. Brown should cease calling on outsiders to advise him as to the electric locomotive's design, without giving me a chance to sit in at the meetings."[45]

Wilgus agreed to the cover-up and ordered his office engineer, Henry A. Stahl, to destroy his only copy of the Woodlawn wreck report, after chief counsel Ira Place reported that President Newman had burned the original in his office fireplace. Wilgus—as well as the company's president and vice-president—knowingly destroyed information crucial to the investigation of the Woodlawn Wreck. Because of the cover-up, the district attorney, state investigators, and the press turned their attention to faulty track maintenance as the most likely cause of the wreck.

Unfortunately for Wilgus, Vice-President Brown did not just suspect the electrification to be the cause of the Woodlawn wreck; he believed Wilgus and the Electric Traction Commission's design of the engine to be at fault. Without Wilgus's knowledge, Brown called in a number of outside railroad engineers to review the design of the new electric engines and propose modifications.

While Wilgus acted in what he perceived to be good faith, he soon learned of Vice-President Brown's meetings with the outside consultants to modify the design of the electric engines. Feeling betrayed, he retaliated by recreating the Woodlawn wreck report from his stenographer's notebooks and the original records. He justified this as a step "to protect my good name in case of need." Quietly, he informed Ira Place of his actions and then took the bold step of sending a copy of the report to the New York Public Library for safekeeping, with the understanding the report would not be available for public scrutiny, without his permission, until he died. Perhaps Wilgus intended the copy of the report in the library as an insurance policy to guarantee that the railroad did not ever attempt again to blame the Woodlawn wreck on his work or that of the Electric Traction Commission.

While Wilgus extracted a degree of revenge by recreating the report and placing it with the library for safekeeping, the events in the spring of 1907 had exerted a toll. That stress, combined with utter exhaustion from his efforts overseeing the Grand Central project, led Wilgus to decide to resign from the railroad and pursue a career as an independent consulting engineer. He explained: "My peace of mind was at stake. . . . I did not wish to remain under circumstance so disturbing to my spirit."[46] Above all else, Wilgus could not have his professional competence challenged by anyone. He submitted his formal letter of resignation to President Newman and the directors of the railroad on July 11, 1907. Newman asked if he would remain to the end of September to tie up loose ends and to ensure that all of the complicated work at 42nd Street would proceed smoothly after his departure.

On September 30, 1907, exactly one year to the day after the first electric engine arrived at Grand Central, Wilgus left his office at the old Grand Central Depot for the last time. He regarded his departure as bittersweet. In his papers he retained testimonial letters from William K. Vanderbilt, J. P. Morgan, Ira Place, and surprisingly, W. C. Brown, whose actions precipitated his resignation. He also included letters of tribute from the engineers who worked for him and his office staff, who gave him an inscribed loving cup in appreciation. Reflecting in later years on his departure, Wilgus tried to measure his accomplishments with the railroad and the Grand Central project: "I dwell in the hope that anyone who may read these lines will not adjudge me to be immodest . . . [and] that I measured up in some degree to the opportunities generously given me to serve the company that employed me and, in so doing, also contributed to the public welfare."[47] Rather modest words from the engineer who envisioned the transformation of a major swath of midtown Manhattan into the most complex transportation and commercial complex in the world.

Completion of Grand Central

At his departure from the New York Central, Wilgus left behind an engineering organization more than capable of completing the construction; and work proceeded on the underground train yard at a frenzied pace. A group of talented engineers and managers picked up exactly where he had left off and carried on with great success. Five more years of effort would be required to complete the underground train yard and the new Grand Central terminal building, which opened formally in February of 1913.

Construction of the underground train yard continued from east to west following Wilgus's "bite" plan. The railroad constructed a temporary terminal building on the east side of the yard at 43rd Street and Lexington Avenue. With

Construction
of Grand Cen-
tral terminal
building, 1909:
erection of
steel trusses
for the roof
of the Grand
Concourse
(looking east
from Madison
Avenue).

the first sections of the underground train yard completed in bite 1 and bite 2, the operating division switched commuter trains to the temporary terminal. Long-distance trains continued to use the old depot until June 5, 1910, when the last express trains departed from the old Grand Central.

Work on the new Grand Central building began in the spring of 1911. A key challenge, as complicated as building the underground train yard, involved the construction of the central space in the building, the Grand Concourse. When completed, the Grand Concourse was the largest interior space in New York City; and because of its innovative engineering, the *Engineering Record* described in detail the erection of the steel frame, including the roof trusses spanning the entire concourse from which the arched ceiling would be suspended.[48]

After a decade of work, the New York Central Railroad scheduled the official opening of the new Grand Central Terminal and underground train yard for the night of February 1, 1913. Reaction in the press to the completion of the project and especially the new terminal building could not have been more laudatory. The popular periodical *Munsey's Magazine* ran a story with the title "The Great-

Table 2.2. Grand Central Terminal versus its rivals, 1913

Railroad station	Total area (acres)	Track length (miles)	No. of tracks	No. of platforms
New Grand Central	70.0	31.8	46	30
Boston South Station	9.2	15.0	32	19
St. Louis Union Station	10.9	5.4	32	16
Paris St. Lazare	11.2	3.5	31	14
Washington Union Station	13.0	14.6	29	13
Pennsylvania Station	28.0	16.0	21	11
Cologne Germany	5.8	3.4	14	9
Chicago & Northwestern	8.0	2.7	16	8
London Waterloo Station	8.75	—	18	—

SOURCE: "New Grand Central Opens Its Doors," *New York Times*, Feb. 2, 1913, sec. 9, p. 1.

est Railroad Terminal in the World" and described the interior of Grand Central in glowing terms as "a new city center; a vast theatre of great events; another triumph of constructive American achievement."[49] Grand Central came to represent more than just a railroad facility; the new terminal stood among the great building projects that transformed New York City into the greatest city in the world. From the opening of the Brooklyn Bridge in 1873, to the first skyscrapers, to the beginning of the largest subway in the world, to the opening of both Pennsylvania Station and Grand Central Terminal, engineers and builders created a new urban landscape unlike any other on earth.

The Sunday issue of the *New York Times* on February 2 devoted a special section to the opening of Grand Central, discussing at great length the splendor and innovative features of the new terminal. The cover page included an artist's rendering of the building with the new viaduct carrying Park Avenue over 42nd Street and around the terminal building. In the distance to the north, the artist created a cityscape of new buildings lining Park Avenue—the new "Terminal City." With the development of the air rights over the underground train yard, the entire complex would cover "an area of thirty city blocks and accommodate 100,000,000 people a year." The new Grand Central would serve as the "newest gateway to New York. . . . Through that gateway in the coming twelve months close to 24,000,000 persons will pass to and from the biggest city in the Western World."[50]

The *Times* also provided a comparison between Grand Central and a number of the world's great railroad stations, both in the United States and abroad (table 2.2). Grand Central, larger than any other railway terminal, had thirty train platforms, eleven more than Boston's South Station, second in number of platforms. No other terminal had a two-story structure with one set of platforms above another, the feature that dramatically increased the capacity of Grand Central. At peak times thirty trains could simultaneously board or discharge passengers; no other rail facility in world boasted anywhere near that capacity.

GRAND CENTRAL'S ENGINEER

Wilgus, writing years after the opening, catalogued a wealth of details about Grand Central's capacity. The miles of platforms on both the upper and lower levels accommodated 559 passenger cars: 392 on the upper level for long-distance trains and 167 on the lower level for suburban service. The 33 miles of tracks in the train yard provided space to store 1,131 passenger cars. The year the new Grand Central opened, the entire rolling stock of the New York Central railroad totaled 1,847 passenger cars, including the electrified cars used for suburban service out of Grand Central.[51] In the two-story underground train yard the railroad provided enough storage capacity for 61 percent of all the passenger cars in service on its vast system of rail lines all the way to Chicago and the Midwest. The terminal building on 42nd Street covered 6.55 acres, and the two concourses totaled 80,844 square feet of space, creating the largest interior spaces in the city.[52]

Midtown Transformed

After the completion of the underground rail yard, an entire new section of New York, "Terminal City," arose, with office buildings, apartments, and hotels built to harmonize with one another. In a 1940 article in *Transactions of the American Society of Civil Engineers*, Wilgus documented the construction, during both phase 1 and 2, of twenty-seven air-rights buildings valued at over $85 million.[53] The first phase of Terminal City involved the construction of a number of buildings adjacent to the new terminal on 42nd Street. Underground passageways connected many of the first air-rights buildings directly to Grand Central and to the upper-level concourse, another distinctive feature. Radiating out from the Grand Concourse a number of corridors, lined with retail shops, connected directly to office buildings and hotels. Wilgus's Grand Central Terminal was the first building complex in the United States to integrate transportation with office, retail, and hotel space. A traveler arriving in New York by train could proceed to a hotel room and later walk a corridor to an office building for a business meeting, perhaps stopping to shop along the way—without ever venturing outdoors. The Biltmore, the fanciest hotel in Terminal City, stood over the incoming station on the west side of the upper level. Arriving passengers rode an elevator from the upper level to the lobby of the Biltmore; porters transferred their luggage from the baggage car of the arriving train directly to their hotel room. Phase 1, from 1908 to 1918, included the construction of twelve air-rights buildings with a total value of over $19 million in 1939–1940 costs (table 2.3).

The second phase of the air-rights development continued through the 1920s till the very eve of the Great Depression. Developers built fifteen more buildings, most along Park Avenue north of Grand Central, valued at over $65 million (table 2.3). The New York Central Building, constructed in 1929, anchored the southern end of the restored Park Avenue, and its two wings included portals through which Park Avenue split around Grand Central via the elevated roadways on each side of the building.

The final building in phase 2, the world-famous Waldorf-Astoria Hotel, replaced a coal-fired power plant originally built to supply electricity and steam

Table 2.3. Terminal City: Buildings constructed as part
of New Grand Central Terminal project, 1908–1931

Building	Address	Year built	Cost (1939–40 $)
PHASE 1, 1908–1918			
U.S. Post Office	Lexington Ave. & 44–45th	1908	$1,800,000
Grand Central Palace	Lexington Ave. & 46–47th	1910	825,000
Adams Express	Lexington Ave. & 45–46th	1914	282,000
American Express	Lexington Ave. & 43–44th	1914	394,000
United Cigar Building	Vanderbilt Ave. & 42–43rd	1914	278,000
Biltmore Hotel	Vanderbilt Ave. & 43–44th	1914	4,600,000
Yale Club	Vanderbilt Ave. & 43rd	1914	725,000
Vanderbilt-Concourse	Vanderbilt Ave. & 44th	1914	850,000
350 Park Ave.	Park Ave. & 51–52nd	1915	675,000
Marguery Hotel	Park Ave. & 47–48th	1917	3,170,000
Chatham Hotel	Park Ave. & 49–50th	1917	950,000
Commodore Hotel	Lexington Ave. & 42–43rd	1918	5,500,000
Subtotal			$20,049,000
PHASE 2, 1920–1931			
460 Lexington Ave.	Lexington Ave. & 45–46th	1920	$3,800,000
290 Park Ave.	Park Ave. & 48–49th	1921	2,325,000
300 Park Ave.	Park Ave. & 49–50th	1921	2,425,000
Knapp Building	Park Ave. & 46–47th	1922	1,325,000
Knapp Building #2	Park Ave. & 46–47th	1922	1,325,000
Park-Lex. Building	Park Ave. & 46–47th	1923	1,850,000
Roosevelt Hotel	Vanderbilt Ave. & 45–46th	1924	4,400,000
Postum Building	Vanderbilt Ave. & 46–47th	1924	3,350,000
277 Park Ave.	Park Ave. & 47–48th	1924	3,400,000
Park Lane Hotel	Park Ave. & 48–49th	1924	2,100,000
Vanderbilt Building	Vanderbilt Ave. & 42–43rd	1925	750,000
Barclay Hotel	Lexington Ave. & 48–49th	1927	2,800,000
Graybar Building	Lexington Ave. & 43–44th	1927	8,500,000
New York Central Bldg.	Park Ave. & 45–46th	1929	10,550,000
Waldorf-Astoria Hotel	Park Ave. & 49–50th	1931	16,200,000
Subtotal			$65,100,000
Total			$85,149,000

SOURCE: William J. Wilgus Papers, Manuscripts Division, New York Public Library, box 4, Research Notes.

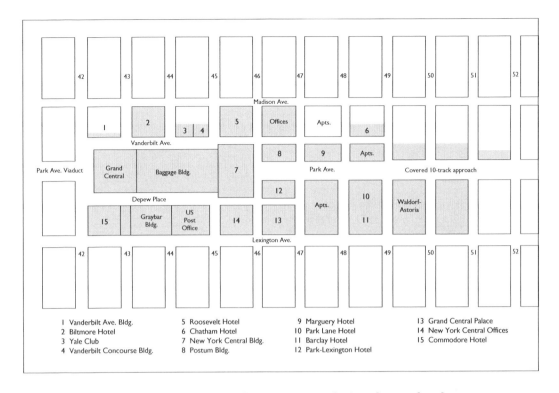

1 Vanderbilt Ave. Bldg.	5 Roosevelt Hotel	9 Marguery Hotel	13 Grand Central Palace
2 Biltmore Hotel	6 Chatham Hotel	10 Park Lane Hotel	14 New York Central Offices
3 Yale Club	7 New York Central Bldg.	11 Barclay Hotel	15 Commodore Hotel
4 Vanderbilt Concourse Bldg.	8 Postum Bldg.	12 Park-Lexington Hotel	

to the new Grand Central and other buildings in Terminal City. The new hotel replaced its predecessor of the same name, constructed by the Astor family in the 1890s on 5th Avenue between 33rd and 34th Streets. Designed by the architects Schultz & Weaver, the new Waldorf Astoria had over fourteen hundred rooms, restaurants and ballrooms, and a residential tower that served the rich and famous.

Grand Central Terminal area, 1933, showing "air rights" development over underground train yard.

In his *Transactions* article, to further illustrate the impact of Terminal City, Wilgus calculated the increase in the assessed value and property taxes in the area immediately around Grand Central and Terminal City between 1904 and 1930. For all of Manhattan, the value of real estate increased by 175 percent during the time period. But in the "Grand Central zone," 42nd Street to 96th, between Lexington and Madison Avenues, property values increased by 374 percent. The dramatic difference was a direct consequence of the completion of the Grand Central project and the elimination of the smoke and steam in the Park Avenue tunnel. Wilgus estimated the difference between actual increase in property values and the values if the zone had increased at the same rate for Manhattan as a whole at half a billion dollars! With the property tax rate of 2.7 percent in 1930, the increased property values generated more than $14 million in additional tax revenue each year for the city of New York.[54]

Grand Central in Perspective

With the completion of Grand Central and Terminal City the east side of midtown Manhattan became a vibrant commercial, residential, and transportation

Table 2.4. Growth in passenger traffic in and out of Grand Central, 1906–1930 (numbers in thousands)

	1906		1930		% increase	
	Commuters	All[a]	Commuters	All[a]	Commuters	All[a]
Hudson Division	3,065	5,829	9,913	14,605	223	151
Harlem Division	3,570	4,905	16,351	19,160	358	291
Total N.Y. Central	6,635	10,734	26,254	33,765	296	215
Total	9,812	19,030	35,708	50,643	264	166

SOURCE: Wilgus, "Grand Central Terminal in Perspective," p. 1020.
[a] Long-distance travelers plus commuters.

center. The new complex's design included a multilevel horizontal and vertical architecture. From the two-story underground platforms and train yard to the twenty-second floor of the Biltmore, to underground corridors radiating outward from the Grand Concourse, a web of connected spaces allowed travelers, office workers, and hotel guests to move about in a world separated from the outdoors. Until the building of Rockefeller Center in the 1930s, no other project created as complex a mix of buildings connected underground. Unlike Rockefeller Center, Grand Central also included the most multifaceted transportation center in the city, if not in the world, with rail connections to the city's suburbs and long-distance service to the Midwest.

The creation of Terminal City served as a catalyst for the transformation of the entire area. Directly across the street on the corner of 42nd Street and Lexington Avenue, Walter Chrysler decided in 1927 to build the tallest building in the world. While the Chrysler Building retained that title for less than a year, the mammoth office building further solidified the emergence of the Grand Central area as one of the city's leading business districts.

Passenger traffic, both long-distance and commuter, grew dramatically from the day the new terminal opened. For the New York Central lines, the total number of passengers increased from just over 10 million in 1906 to over 33 million in 1930, an increase of 215 percent (table 2.4). The really astounding growth occurred in the suburban commuter service. In 1906, of the 10.7 million passengers the New York Central carried in and out of Grand Central, commuters accounted for 62 percent. By 1930, commuters constituted 78 percent of the Grand Central passengers on the New York Central trains; the number of commuters increased by 296 percent during the time period. By 1930 the vast majority of passengers using Grand Central were not riding the glamorous long-distance trains to Buffalo, Cleveland, and Chicago; rather, they were rushing in the evening to catch the 5:22 to Scarsdale or the 5:53 to Larchmont.

When Wilgus first presented his overall plan for Grand Central to the New York Central executives and board of directors in 1902, he argued that the improvements, especially the electrification eliminating the awful steam and smoke in the Park Avenue tunnel, would lead to increased suburban traffic. By 1930, suburban commuters exceeded the number of long-distance passengers by three to one, despite being relegated to the lower-level concourse.

From every aspect—architectural, technological, operational, and the impact on the civic life of midtown Manhattan—the new Grand Central and Terminal City succeeded beyond even the most optimistic projections. In summarizing the project, a milestone in his own career, Wilgus drew a link from the new terminal back to Cornelius Vanderbilt: "In the mind of Commodore Vanderbilt was born the idea of a new terminal . . . to which was to be brought the traffic of a great railroad system under his control." The result: "The inescapable substitution of electricity for steam . . . combined with the idea, born in 1902, that revenue plucked from the air might be used to finance its tremendous cost. . . . The project was rounded out and blessings reaped . . . in the heart of the City of New York . . . a remarkable civic center."[55] While Wilgus clearly had a vested interest in glorifying the work that came from his own accomplishments, the usually conservative and objective *Scientific American* published a lead article anticipating the completion of the Grand Central project entitled "The World's Greatest Railway Terminal."[56]

Departure

With some bitterness, Wilgus admitted in his memoirs that at one time he had hoped to lead the New York Central, but by the time he left the railroad, "I had ceased in my heart to aspire to eventual promotion to post of president of the railroad company." He added, "I preferred that my future should be spent in a wider field giving sway to my inherently independent spirit."[57]

Wilgus knew that leaving the New York Central Railroad would have consequences for his reputation, especially in regard to the key role he had played in the Grand Central project. After he successfully defended the design of the new electric engines, blame for the Woodlawn wreck shifted to the operating division, headed by W. C. Smith. Wilgus remained convinced that Smith had played a crucial role in downplaying Wilgus's contributions to the railroad: "Down to the present time a New York Central policy, springing from my falling out with Mr. Smith, has had its purpose realized in the casting of my doings and those of my old associates into the background, our names, in so far as the company is concerned, into undeserved oblivion."[58] That "oblivion" included no mention of Wilgus at the opening ceremonies for Grand Central in 1913 or even the inclusion of his name on the official brochure.

Wilgus was not to remain idle. An old friend from his days working in upstate New York, Henry J. Pierce, the former president of the street railroad company in Buffalo, New York, invited him to join him in a consulting company, the Amsterdam Corporation. As soon as Wilgus joined Pierce, they set out to develop plans to solve the New York Central's long-running "west side problem"—the jumble of tracks and freight rail yards running along the Hudson River from the upper west side to the Battery.

3

NEW YORK'S

FREIGHT

PROBLEM

AFTER DRAMATICALLY RESIGNING from the New York Central Railroad in September of 1907, Wilgus began a second career as engineering consultant. With an old friend, Henry J. Pierce, also a railroad engineer and executive with whom Wilgus had worked in Buffalo, New York, while still with the Central, he formed a consulting business. Their new firm, the Amsterdam Corporation, rented office space at 165 Broadway in lower Manhattan. A major opportunity presented itself as soon as Wilgus and Pierce went into business: the need to alleviate the railroad and freight congestion along the Hudson River.

For decades New York City politicians, residents, and the press demanded a solution to the railroad and shipping problems on Manhattan's west side. Simply referred to as the "west side problem," the transportation nightmare was created by a combination of shipping piers, ferry slips, and railroad facilities along the Hudson River from 60th Street south to the Battery. As we have already seen, thousands of ships berthed each year at the Hudson River piers; and ten passenger and freight ferries occupied piers on the river between Manhattan Island and the New Jersey side of the harbor, where both people and goods were unloaded and loaded for the trips back and forth across the river. The thousands of tons of freight delivered to the shore had to be hauled to final destinations on the island, often creating gridlock on the streets and avenues in lower Manhattan. Added to the "local" traffic from the hinterlands were the tons of goods involved in overseas import and export, as well as the coastal shipping to Southern and New England ports.

To compound the chaos, the New York Central & Hudson River Railroad owned track rights from 60th Street all the way down to St. James Park, just north of the Battery. For a distance of over five miles, the railroad possessed the legal right to use the city's streets for its railroad tracks. Each day steam engines hauled scores of trains and thousands of freight cars back and forth between 60th Street and St. James Park. Pedestrians took their lives in their hands as they crossed the tracks. Even with a railroad employee riding a horse and waving a large red warning flag to announce a train, numbers of city residents were killed each year by passing freight trains. Newspapers referred to the streets on the west side, especially those along 10th Avenue, as "Death Alley."

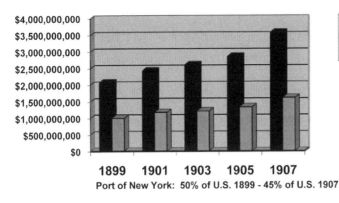

Foreign commerce of the United States, 1899–1907: exports and imports.

With a population of over 1.8 million residents in Manhattan alone in 1900 and over 3.4 million in the expanded city of New York—a result of the consolidation with the four other boroughs—the city needed to be supplied daily with an enormous amount of freight. Coal, for example, provided not just heat in the winter but also illumination. All of the early gas companies burned coal to manufacture gas to be distributed for lighting, and the first electric companies burned coal to produce electricity. The city's transportation chaos did not end on the shores of Manhattan. Once the freight arrived at the piers, horse-drawn wagons—drays—hauled it to thousands of locations throughout the city, especially to the manufacturers and businesses in lower Manhattan. Wilgus estimated that over 23 million tons were delivered in this fashion in 1906–7 (table 3.1).

A reverse flow of freight only added to the congestion. Manufacturing companies in Manhattan shipped their products all over the world. Drays carried the merchandise to the piers to be loaded onto ships bound for ports around the world or along the U.S. coast. Other freight had to be loaded onto car floats or lighters and hauled back across the Hudson to New Jersey and then loaded onto departing freight trains for distribution throughout the United States. Even though Manhattan had less than 5 percent of the country's population, its manufacturing firms accounted for a major share of total U.S. manufacturing from 1880 to 1900 (table 3.2).

On a more mundane level, all of the coal burned in New York created ashes to be removed along with the city's garbage. And perhaps as many as 100,000 horses left tons of manure on the streets to be loaded onto barges for disposal.

In short, nowhere else in the entire country did any port have a more complicated transportation system than Manhattan. On one small island, even an island that claimed to be the center of the world, the surrounding rivers and bays created an obstacle to easy communication and transport. And while attention focused on the "west side" problem, the East River shore also teemed with piers, shipping, ferries, and barges. New York City's transportation challenges in the early 1900s remained as difficult to solve as when Cornelius Vanderbilt had begun his modest ferry service from Staten Island to Manhattan almost a hundred years earlier.

An example of the "west side problem": a New York Central freight train on 11th Avenue, 1910.

Table 3.1. Freight requiring drayage in Manhattan, 1906–1907 (thousands of tons)

Carrier	Railroad & foreign line merchandise	All merchandise	Coal	Total
Railroads				
Rail delivery via N.Y. Central	2,000			
Water delivery	7,800	9,800	4,000	13,800
Foreign steamship lines				
Terminating in Manhattan	1,400			
Terminating in New Jersey	800	2,200		2,200
Coastwise lines		2,900		2,900
River and Long Island Sound lines		1,300		1,300
Irregular lines		3,000		3,000
Total		19,200	4,000	23,200

SOURCE: Wilgus, "Plan of Proposed New Railway System for the Transportation and Distribution of Freight by Improved Methods," William J. Wilgus Papers, Manuscripts Division, New York Public Library, box 45, 12.

Origins of the West Side Problem

The transportation bottleneck on the west side of Manhattan began in 1846 at the dawn of the railway age in the United States. In May 1846 the New York legislature chartered the Hudson River Railroad with authority to construct a rail line from Poughkeepsie, New York, to New York City, along the Hudson River.[1] The charter also granted the Hudson River Railroad the right to extend its tracks north of Poughkeepsie to Albany. Crucially, the legislation gave the

Table 3.2. U.S. Census of Manufactures, 1880, 1890, 1900

	1880			1890			1900		
	U.S.	New York City	N.Y. as % of U.S.	U.S.	New York City	N.Y. as % of U.S.	U.S.	New York City	N.Y. as % of U.S.
No. of establishments	253,850	16,923	7	355,405	37,163	10	512,276	39,776	8
Capital	$2,790	$251	9	$8,525	$622	9	$9,831	$921	9
Employees	2,732,595	272,920	10	4,251,535	415,552	10	5,314,539	462,763	9
Wages	$947	$120	13	$1,891	$236	13	$2,327	$245	11
Cost of materials	$3,396	$423	12	$5,162	$538	10	$7,346	$709	10
Value of products	$5,369	$659	12	$9,372	$1,084	12	$13,010	$1,317	11

SOURCE: Census data reported in Wilgus, "Plan of Proposed New Railway System," 6.
NOTE: Dollars in millions.

railroad the right to use the streets in the city of New York for its tracks, with the approval of the city: the Hudson could "locate [its] railroad on any of the streets or avenues of the city of New York westerly of and including the Eighth Avenue, and on or westerly of Hudson Street, provided the consent of the corporation of said city be first obtained."[2]

Without any hesitation the city of New York passed legislation in May of the following year allowing the Hudson River Railroad to construct tracks from Spuyten Duyvil down the west side of Manhattan to near the Battery at the tip of the island. The act specified the railroad's right to build its tracks on 12th, 11th, and 10th Avenues and on West, Hudson, and Canal Streets to the southern end of the island. With the passage of the enabling legislation in 1847, the city of New York gave the Hudson River company a franchise of immense value: the right to bring its trains onto Manhattan Island. Only the then still independent New York & Harlem Railroad held a similar franchise, with the right to use 4th Avenue on the east side for its tracks.

The Hudson River Railroad's charter, however, also included a clause prohibiting the railroad's tracks on the city streets from interfering with other ordinary uses of the streets by residents and businesses. Here lies the source of conflict between the railroad, the city government, and ordinary citizens that continued for over eighty years and led to endless debate, anger, and litigation for all parties involved.

In the next chapter of the railroad's history Commodore Vanderbilt enters the story. By 1869 he had gained control of the Hudson River Railroad. Since he already controlled the New York Central, Vanderbilt decided to merge his two railroads to create one railroad system with tracks stretching from Buffalo across New York State to Albany, and then from Albany down to New York and along the west side of Manhattan to the Battery. Vanderbilt also already controlled the New York & Harlem and wanted to merge all three railroads. A battle royal ensued in Albany, with Vanderbilt hauling bags of money to "persuade" reluctant state legislators to approve his grand merger plans. For once, Vanderbilt did not get his way with the legislature. The law passed in 1869 allowed him to merge only the New York Central with the Hudson River.[3] Since Vanderbilt could not formally merge the New York & Harlem with his other railroads, he had to settle for coordinating the running of the Harlem with the newly formed New York Central & Hudson River Railroad. With the merger, the rights to use the city's streets on the west side passed to the Central railroad.

Vanderbilt now effectively controlled all direct railroad access to Manhattan Island, a franchise of immense value. His railroads held a monopoly on both passenger and freight service to New York City. For the next forty-three years, until the Pennsylvania Railroad battled its way onto Manhattan Island with a passenger tunnel under the Hudson River, no other railroad enjoyed direct access for passengers and freight, ensuring the Central a perpetual competitive advantage. Vanderbilt lived for only eight years after the merger, but his son William Henry and his grandsons Cornelius III and William K. exercised control over the New York Central and fiercely defended the railroad's use of the city's streets.

When the Hudson River Railroad first received its charter in 1846 and constructed rail lines down the west side of Manhattan, the city's population totaled just over three hundred thousand, with most of the population crowded into the lower third of the island, below 14th Street. At first the railroad's track along the Hudson River did not interfere with the city's residential or commercial activity. But as the population increased, the settled areas of the city moved northward, and the Hudson River's tracks become surrounded by crowded streets and neighborhoods. Inevitably, the dual use of the city avenues for railroad purposes and city life clashed. As a New York State investigation concluded, train operations on the west side "became a steadily growing menace to life and limb and property, and Tenth Avenue became notorious as 'Death Avenue.'"[4]

The Hudson River Railroad constructed a major depot in lower Manhattan at St. John's Park and Varick Street. The St. John's depot served both passengers and freight until 1871, when Commodore Vanderbilt opened the first Grand Central and switched all of the Hudson River passenger trains to 42nd Street. Freight service continued to St. John's depot, where horse-drawn drays picked up the freight for final delivery. The Hudson's tracks along the river provided for a lucrative business hauling freight to the piers lining the river.

An exhaustive study completed by the immediate predecessor of the Port Authority of New York reported that the New York Central & Hudson River Railroad handled 1,672,445 tons of freight below 60th Street in 1914. To move that volume of freight required the railroad to move a total of 270,988 freight cars back and forth.[5] If the average freight train included ten cars, thousands of trains a year, and perhaps hundreds a day, moved up and down the railroad's tracks. To compound matters, the steam engines that pulled the trains added smoke, steam, soot, and noise to the bedlam along the river.

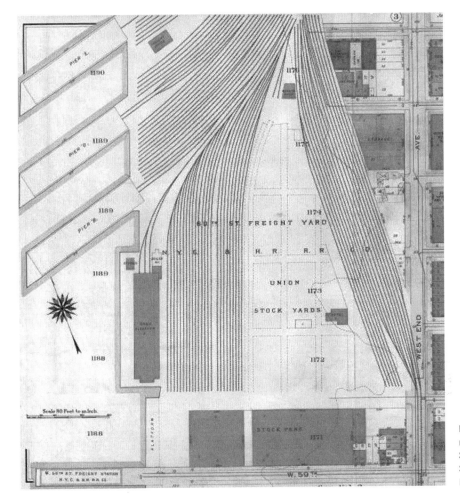

New York Central rail yard, 65th Street to 59th Street and Hudson River.

The Rail Yards at 60th and 30th Streets

In addition to the terminal at St. John's Park in lower Manhattan, the Hudson River Railroad owned two large rail yards on the west side at 60th and 30th Streets. The railroad had constructed both yards before development had reached the areas around 30th and 60th Streets. The 60th Street yard stretched north to beyond 71st Street, and from 11th Avenue to the Hudson River. This facility provided direct access for the railroad's freight cars to the piers on the river. Next to the river at 60th Street, the railroad built an enormous grain depot. The grain elevators stored wheat shipped from upstate New York and the Midwest that was to be loaded on ships for export. The Union Stock Yard occupied the blocks between the river and West End Avenue. The railroad delivered live cattle, also shipped from the Midwest, to be slaughtered, providing New York with fresh meat, a necessity before the advent of reliable refrigeration.

At 30th Street the railroad owned a second rail yard with adjacent piers on the Hudson. After the Hudson railroad merged with the New York Central and the Vanderbilts expanded their railroad empire, they added the West Shore Rail-

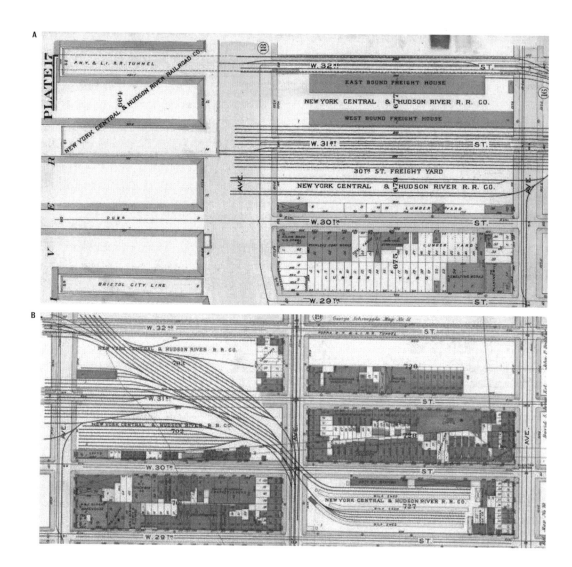

New York Central rail yard: (A) 32nd Street to 29th Street and Hudson River; (B) 32nd Street to 29th Street between 9th and 11th Avenues.

road, with tracks on the west side of the Hudson, to Albany and beyond. Freight shipped on the West Shore and destined for Manhattan arrived on the New Jersey side of the river at Weehawken. Tugs ferried the freight cars, loaded onto car floats, across the river to the 30th Street rail piers. Further inland between 9th and 10th Avenues, a huge New York Central & Hudson River Railroad freight depot occupied the entire block between 29th and 30th Streets. With tracks to the depot crossing both avenues at street level, freight cars periodically blocked all north-south street traffic. Tracks on 10th Avenue, "Death Alley," continued south from the 30th Street yard, eventually winding their way down to lower Manhattan.

The two rail yards on the west side provided the only freight facilities on Manhattan Island. Along with Grand Central and track rights down Park Avenue to 42nd Street for passenger trains, the New York Central maintained a monopoly on direct rail access for both passengers and freight. While the officers of the railroad and the Vanderbilts often spoke of their dedication to serving the citizens of New York City, they steadfastly refused efforts to resolve the congestion and hazards caused by their freight tracks occupying the city streets from the very northern tip of Manhattan to St. John's Park with two major rail yards at 30th and 60th Streets. From the first legislation chartering the Hudson River Railroad in 1846 through the merger of the Hudson with the New York Central in 1869, the railroad viewed the right to use the streets of New York, and to build and later expand the rail yards at 30th and 60th Streets, as an absolute legal right. Even as the population of the city grew and the west side developed, the railroad vigorously defended its use of city streets for freight trains.

In 1917, the New York State Legislature appointed a commission to once again investigate the West Side problem. The commission pointed out that both the New York Central and the Pennsylvania railroads had spent enormous sums on their new passenger terminals in New York—Grand Central Terminal and Pennsylvania Station. When added to the money invested in the city's first subway lines and new elevated railroads, "approximately $700,000,000 has been expended in various ways in the improvement of passenger transportation and scarcely a dollar to improve the facilities for transporting quickly and economically the things produced by manufactures or required in daily life." In the opinion of the commission the Central had "led many to think that it is more deeply

A New York Central freight train on West Street along the Hudson River, circa 1910, looking north from 30th Street.

interested in maintaining its existing claims [to use the city streets on the west side] than in the development of freight facilities."[6]

This 1917 report echoed a rising chorus of public complaints about the Central tracks on the west side of the island. A decade earlier, political pressure had led the New York State Legislature to pass a draconian law in 1906, known as the Saxe Law, requiring the Central to end the use of the streets of New York.[7] The act authorized the Rapid Transit Commission to negotiate with the railroad to remove its tracks or face losing the right to operate on the city's streets. Needless to say, the railroad challenged the legislature's move to abrogate its legal right, under the charter of 1846, to use the city's streets. To be even more assertive the railroad claimed that when the state legislature approved the merger of the Hudson and Central in 1869 for a period of "500 years," it implicitly extended the railroad's track rights for the same time period; five hundred years meant in perpetuity.

Direct rail access to Manhattan had proven to be priceless, and the railroad stood ready to defend its rights vigorously. Ira Place, the New York Central's chief legal counsel, when called to testify before a state commission argued that the legislature had no right to order the railroad to remove its tracks, since the Hudson's original 1846 charter was extended for five hundred years when the Legislature approved the merger of the Hudson and the Central in 1869. When questioned, Place defended the railroad's rights:

> Q. In 1847, there was an ordinance [New York City] adopted assenting or consenting to the acts of the railroad under the charter of the road of 1846 [New York State]. Is that correct?
>
> Place A. Yes. . . .
>
> Q. What did you do below 48th street?
>
> Place A. We constructed the road to 30th Street under the ordinance and down to Chambers street, under the ordinance of 1846. . . .
>
> Q. Granted by the ordinance?
>
> Place A. They came from the State. We got certain rights from the State. The State gave us the right to build where we did build [on the city streets] under the condition that we got the consent of and assent of the city and when we got it, it was fixed. . . . The charter and the acceptance of the charter, and the building of the road, and the condition imposed of obtaining the consent and assent of the city having been made. That constituted our rights.[8]

Later in his testimony the railroad's chief counsel pointed to an appellate court decision upholding the railroad's right to use the city streets and added, "The court said we had these rights, and we have them now." The railroad stated its willingness to work with the city to alleviate the worst of the problems on the west side, but the Central would not budge an inch in claiming a perpetual right to maintain its freight tracks along the Hudson River.

But the west side problem involved more than just the New York Central

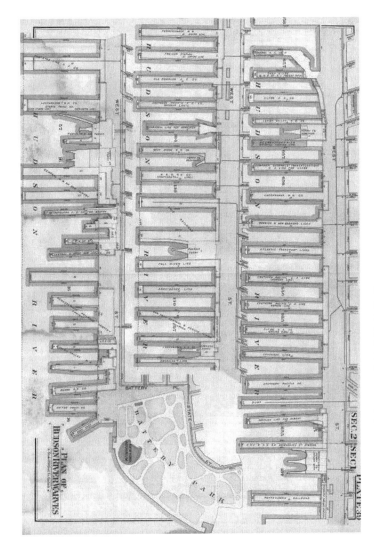

Hudson River piers along West Street: *left*, pier 1 at Battery to pier 14, Fall River Line, at Fulton Street; *center*, pier 15, Providence Line at Vesey Street to pier 28, Pennsylvania Railroad at Laight Street; *right*, pier 29, Pennsylvania Railroad to pier 47, Quebec Line at Charles Street.

freight service. Along the Hudson River, piers lined the shore from the Battery north to the end of Central's rail yard at 60th Street. Shipping companies, railroads, and ferries all either owned or leased piers on a long-term basis with no rhyme or reason except historical happenstance for occupying a particular pier.

From 72nd Street south, piers lined the river for almost six solid miles. Wilgus, on behalf of his consulting company, obtained information from the New York Department of Docks detailing the usage of the Hudson waterfront. His analysis found that the steamship companies engaged in foreign commerce occupied only 23 percent of the piers and frontage along the "marginal way," the term used for the land along the river (table 3.3). By comparison the railroad's facilities for their car floats, freight lighters, and barges took up a much larger share of the piers and pier frontage. These facilities served the railroads whose tracks ended across the harbor in New Jersey and had to float their freight across the harbor to Manhattan. Ten ferries, carrying both passengers and freight,

Table 3.3. Piers and frontage along Hudson River, 1907

User	No. of piers	Lineal feet	% of total marginal way
Railroads	8	9,460	30
Foreign shippers	12	7,446	23
Coastwise & Long Island Sound shippers	16	4,084	13
Ferries	10	2,132	7
Municipal	5	1,962	6
Miscellaneous	—	6,451	21
All users		31,535	

SOURCE: Wilgus, "Plan of Proposed New Railway System," 11.

operated between Manhattan and New Jersey; coastwise and Long Island Sound shipping occupied additional space. All of the freight arriving by ferry, car floats, or lighters had to be hauled away from the piers along the waterfront. Hundreds of horse-drawn drays jammed the streets in front of the piers waiting a turn to load or unload.

While the press and public officials focused attention on the New York Central's tracks and trains when debating the west side problem, the amount of freight delivered to Manhattan by the shipping companies, ferries, and the other railroads dwarfed the freight carried by the New York Central. Wilgus estimated the total tonnage of merchandise delivered to Manhattan in 1907 at 19.2 million tons; of that total, just 2 million tons arrived on Manhattan via the New York Central.[9]

Just north of the Battery, Pennsylvania Railroad freight occupied pier 1, while the Lehigh Valley Railroad operated a freight terminal on piers 2 and 3. Piers 4 and 5, with a large freight building on the pier front, served as the Pennsylvania Railroad's main Manhattan freight warehouse. Further north, the Pennsylvania Railroad's passenger ferry landed at the pier across from Courtland Street. Until the opening of Pennsylvania Station in 1912, all of the company's passenger trains terminated in Jersey City and passengers rode the ferry across the river to the ferry building to complete their trip to New York. Scores of horse-drawn cabs arrived and departed at all hours of the day and night carrying passengers on the last leg of their journey to New York City. Two "Sound" lines, the Providence Line and Fall River Line, occupied piers 18 and 19. Both lines offered passenger and freight service to New England via Narragansett Bay; the Fall River Line ships stopped at Newport, providing luxury service for wealthy travelers who summered in Newport.

Foreign shippers also occupied numerous Hudson River piers. Just a little north of Spring Street, the Clyde Steamship Line offered service to England from pier 36, and on the next pier the Southern Pacific Line sailed between New York and the Orient. At pier 42, Morton Street, the Compagnie General Transatlantic ran its freight and passenger ships between New York and Le Havre, France. The Quebec Steamship company used pier 47 at Charles Street.

Just as there was no logical pattern in use of the piers along the Hudson, neither was there on the East River and on the Brooklyn waterfronts. As the port grew and more shipping companies and railroads established facilities, they occupied any available pier. The waterways filled with ships, ferries, barges, car floats, and tugboats hauling freight and passengers to myriad locations. Sheer chaos reigned, and a solution was desperately needed.

Previous Plans to Solve the West Side Problem

As early as 1867, railroad engineers had offered a series of plans to deal with traffic congestion along the Hudson. Some advocated moving the railroad tracks that were using 10th and 11th Avenues westward to the river's edge. On city land in front of the piers lining the river—land referred to as the "marginal way" and later as West Street—William Bryant, a prominent civil engineer, proposed building a four-track waterfront railway. In 1874, several members of the American Society of Civil Engineers proposed building an elevated freight railway over West Street from Chambers Street in lower Manhattan to 23rd Street. At the turn of the century, William Parsons, chief engineer, prepared a study for the Rapid Transit Commission and also suggested an elevated freight railway along the marginal way. Calvin Tomkins, president of the Municipal Art Society and later commissioner of docks, in 1904 offered a plan for a freight subway along the west side and an elevated railway for passengers.[10] The New York Central and the city of New York discussed a plan for the railroad to build a subway for its tracks between the 60th and 30th Street rail yards, eliminating the tracks on 10th Avenue for a distance of thirty blocks.

All of these plans focused on only a part of the problem: the Central's railroad tracks on the city's streets, and the movement of its freight up and down the island between St. John's Park Depot and the 30th and 60th Street rail yards. None of these plans addressed the larger problem of freight delivered to Manhattan by car floats and lighters from the New Jersey side of the harbor. An elevated railway or a freight subway would relieve only some of the congestion. The chaos created by freight delivered across the harbor to the piers would not be mitigated in the least. The ten railroads on the Jersey side of the harbor would continue to float their freight to and from Manhattan to be hauled from the piers, on the city streets, to final destinations.

The Public Service Commission and William Wilgus

The Saxe Law, passed by the New York State Legislature in 1906, directed the Rapid Transit Commission to oversee the removal of the Central's tracks on the west side. The long, complicated story of efforts to bring rapid transit to New York evolved over decades and included a fascinating cast of characters.[11] By 1906 Manhattan had four major elevated railroads with steam-powered locomotives as well as the first subway, the IRT (Interborough Rapid Transit), powered by electricity. The Rapid Transit Commission, established in 1894, achieved success but came to be regarded by the public and business interests in New York as out of step. Political pressure led the state legislature in 1907 to replace the Rapid Transit Commission with a new regulatory body, the Public Service Commission

(PSC), with two districts: the First District for New York City and the Second District for the rest of the state. Unlike the Rapid Transit Commission, which had included the mayor of New York and the city controller as *ex officio* members, the Commission for the First District consisted of five members appointed by the governor. Politics never seemed to be far removed from any efforts involving transportation in New York City or the metropolitan region. Supporters of the PSC argued that the new commissioners would be more professional and bring a nonpartisan point of view to crucial decisions involving future transportation improvements.

William Wilgus's reputation among professional transportation engineers remained high despite his recent resignation from the New York Central Railroad. In late March 1908, Edward M. Bassett, one of the PSC commissioners, asked Wilgus and his Amsterdam Corporation to serve as a consultants to investigate the west side problem. New York City and the Central Railroad remained deadlocked, and the commission wanted a fresh perspective on solving the transportation chaos along the shore of the Hudson River. In a letter to Wilgus, William Willcox, another commissioner, encouraged Wilgus to think boldly: "Make your report cover the entire field of information that should furnish a basis for the right solution of the west side problem."[12] Never one to need encouragement to think boldly, Wilgus immediately accepted the offer to prepare a detailed analysis and offer solutions to the myriad problems on Manhattan's west side.

In April, Wilgus sent a formal proposal to the commission outlining a two-step process. A first report would analyze "the freight traffic problems of lower Manhattan Island from the standpoint of the City at large, the railroads, and the shipping interest" for a consulting fee of $5,000. A second report, proposing a solution to the west side problem, would be prepared for an additional fee to be negotiated later.[13] The PSC approved the contract with the Amsterdam Corporation on May 3, 1908, and Wilgus's long career as a private consultant commenced with his proposal to the commission.

Wilgus's Analysis of the Freight Problem

In his usual methodical manner, Wilgus set out to analyze the complicated west side freight problem. From the very beginning, he realized the problem did not involve just the New York Central tracks and rail yards along the Hudson River. While the New York Central had an advantage over all of the other railroads serving the port of New York because of its direct access to Manhattan Island, the Central's share of freight delivered in the city accounted for just 20 percent of all freight distributed by the railroads (table 3.1). Removing the railroad's tracks from 10th and 11th Avenues and replacing them with a subway or elevated railway would not solve the problem. Any solution had to include improved delivery of freight from the railroads terminating on the New Jersey side of the harbor.

Wilgus first gathered data and maps to begin the analysis. He listed the materials needed and set his staff to work collecting the information:

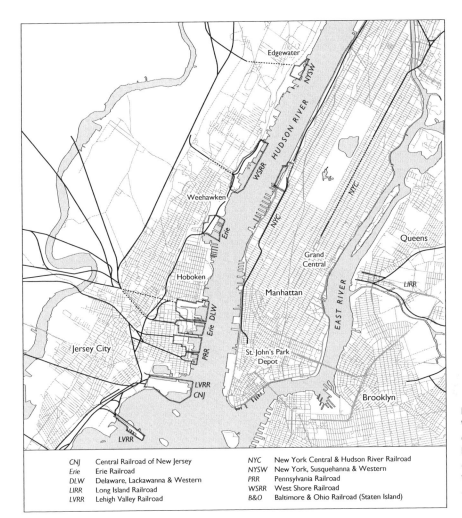

New Jersey
Waterfront, 1910:
eight railroads
had terminals
and rail yards on
the New Jersey
side of New York
harbor.

CNJ	Central Railroad of New Jersey	NYC	New York Central & Hudson River Railroad
Erie	Erie Railroad	NYSW	New York, Susquehanna & Western
DLW	Delaware, Lackawanna & Western	PRR	Pennsylvania Railroad
LIRR	Long Island Railroad	WSRR	West Shore Railroad
LVRR	Lehigh Valley Railroad	B&O	Baltimore & Ohio Railroad (Staten Island)

1. Maps of waterfront showing widths of marginal streets and usage of the various waterfront piers
2. Location of the principal large shipping establishments, bonded warehouses and either places from which large quantities of freight are received or shipped in lower Manhattan
3. Map showing location of all railroads carrying freight terminating on the west side of the North River [the Hudson]
4. Photographs showing typical congestion of drays on West Street and also in another representative congested location in lower Manhattan
5. Cost per ton of handling freight from car to freight house and freight house to dray
6. Lineal feet of waterfront within what may be termed "the busy portion of Manhattan Island," say south of 72nd St. North River and south of 42nd St. East River, portions used by—city of New York, Railroads, Steamships, others
7. Tonnage and cost of handling same, per blank attached to . . . the railroads[14]

The last item referred to letters sent to all of the railroads asking each to provide detailed data on the amount of freight carried into and out of the port of New York and the associated costs of transportation. Wilgus immediately encountered difficulty in obtaining the needed information on freight tonnages. The railroads regarded the data, especially costs, as proprietary and did not want their rivals to have access to this information. Wilgus assured each of the railroads that he would keep their individual figures confidential and report only aggregate data.

As the research on freight costs continued, Wilgus realized that the railroads on the New Jersey side of the Hudson provided merchants and manufacturers on Manhattan Island and in Brooklyn with a sizable hidden discount. It cost each railroad a substantial sum to move freight from its Jersey rail yards across the harbor and then on to the city waterfront. But the railroads did not charge extra for this service (referred to as "lighterage") because of the competition from the New York Central; all of the railroads charged approximately the same fees to deliver a ton of freight to lower Manhattan. With the only rail tracks on the island, the Central did not face the additional costs of hauling freight across the harbor. No wonder Ira Place, the chief counsel of the Central who testified before the state commission in 1917, vigorously defended the railroad's right to continue freight service on the west side of the island.

Wilgus reasoned that if the state and city forced the Central to remove its tracks from the streets on the west side, effectively curtailing the railroad's freight service to lower Manhattan, the other railroads would immediately add the costs of lighterage to their freight charges. To test his idea about the potential for increased costs, he sounded out Percy Todd, vice president of the Bangor & Aroostook Railroad, an old friend and previously a senior manager with the West Shore Railroad, a subsidiary of the Central. In a letter to Todd dated May 5, 1908, Wilgus said: "It has been my understanding that due to the presence of the New York Central on the west side of lower Manhattan, all of the railroads terminating on the west side of the North River absorb their lighterage charges," and then explained his supposition—"so that, should the New York Central tracks be abolished, the merchants and manufacturers on Manhattan Island face the possibility or even probability of addition of lighterage charges."[15] Todd agreed completely: "There is no doubt in my mind whatever that the original cause of the roads terminating on the west side of the Hudson River making lighterage free arrangements around the harbor was due to the fact that the New York Central reached St. John's Park. . . . There is no idea in my mind but what a strong effort would be made, if the St. John's Park line to be abandoned, to abolish all free lighterage."[16]

For Wilgus the costs to the New Jersey railroads of "free" lighterage provided the key to his solution for the west side problem. What if the railroads could transport the freight directly onto Manhattan Island via a shared underground freight subway under the Hudson River and then down through lower Manhattan under the city streets? His calculations showed the charges incurred for using the new freight subway to be substantially less than the costs of lighterage.

The railroads would both save money and benefit from a vastly more efficient system of delivering freight to and from Manhattan Island. As with the Grand Central project, Wilgus, in a stroke of genius, conceived of an innovative plan to solve the entire freight problem in the port of New York. He hinted to Todd of his radical plan: "As a result of the investigation up to this point I am seriously considering to suggest to the Public Service Commission of an entirely novel method of handling freight from the railroad terminals direct to the shipper without the intermediary costs of lighterage and drayage."[17]

Wilgus also alerted Bassett and the PSC: "I think that you will be interested in knowing that I have the Manhattan freight situation well in hand, and . . . among other possible solutions have a novel one."[18] As with his Grand Central plan, Wilgus focused on the entire scope of the problem—the transportation system as a whole and not just one isolated component. The "freight situation" involved hundreds of private shippers, steamship lines, twelve railroads (including the two largest in the country), and millions of dollars of investments in piers, ferries, lighters, barges, and tugboats. To Wilgus the freight transportation problem involved the entire port of New York and not just the Central's tracks on 10th Avenue. With such a complex problem, only a regional, integrated approach offered a real solution.

The Wilgus Plan: A Small-Car Freight Subway

Throughout the spring and summer of 1908 Wilgus continued to work on the study for the Public Service Commission, including a proposed solution: a small-car freight subway. The consulting contract called for the report to be delivered to the commission in September 1908.

The plan proposed building a subway tunnel under the Hudson River and a belt railway in New Jersey linking all of the railroads on the west side of the Harbor. A tunnel through Bergen Hill would connect the subway tunnel to a new rail yard in the Hackensack Meadows serving all of the railroads. In Manhattan, a freight subway would be built around the shore of the lower end of the Island and then north under 1st Avenue to link with the railroads in the Bronx. Throughout the business district in lower Manhattan, small-car subway lines would run under the city sidewalks, providing underground delivery directly to piers, freight terminals, factories, and stores at basement level. Of course, the subway would be electric-powered.

In the fall Wilgus testified before the PSC and presented the major components of the plan:

a) *New Jersey Belt Line:* A belt line in New Jersey extending from the Baltimore & Ohio R.R. on the south to the West Shore R.R. on the north, a distance of about 15 miles. . . . At each intersection with these trunk lines interchange facilities are proposed for economically and quickly transferring the freight between the standard large cars and small cars of the belt line. . . . Small cars shall be assembled and classified for delivery to their respective destinations in a large gravity yard just west of Bergen Hill.

b) *Manhattan Belt Line:* From the classification yard the small cars are to traverse a four track tunnel beneath Bergen Hill and the North River to a connection with a high-speed belt line circumscribing the shores of Manhattan Island from the 60th Street yard of the New York Central on the North River to the Battery, and hence along the East River and up First Avenue to a connection with the railroads in the Bronx. A cross-town line is also proposed at some convenient point north of 23rd Street, connecting the North and East River lines with the route to New Jersey

c) *Sidewalk Subways:* From the Manhattan belt line, subways are proposed beneath the sidewalks, tapping both sides of the major wholesale streets below 42nd Street and connecting the important wholesale and retail districts of the middle portions of the Island.

d) *Transfer Stations:* For the portion of traffic that cannot be handled directly through the sidewalk subways to the shippers, transfer stations with overhead storage facilities . . . are to be provided at frequent intervals.[19]

At the transfer stations, freight cars destined for locations not directly served by the subway would be lifted off their wheels directly onto flat-bed delivery trucks. This ingenious transfer system, anticipating the freight container by fifty years, eliminated the need to unload the freight at the stations.

Impressive in scope, the plan envisioned handling 90 percent of the freight then delivered to lower Manhattan by all of the railroads serving the port of New York. With freight handling removed from the New Jersey shore and the piers on the Hudson and East Rivers, valuable pier space could be converted to shipping instead of railroad use. The delivery of freight by subway under the city's sidewalks directly to the cellars of the commercial buildings would eliminate the congestion on the city streets caused by thousands of drays moving freight throughout lower Manhattan. Wilgus planned the small-car subway to run under city sidewalks and not interfere with the passenger subways constructed under the adjacent city streets.

Wilgus certainly seemed pleased with himself and his company's bold plans to completely reorganize the delivery of freight to and from Manhattan Island. Just over one year after leaving the employ of the New York Central, with the Grand Central project still far from completion, Wilgus had proposed a plan as intrepid as Grand Central, but this time to solve New York's freight transportation issues and eliminate the "west side problem." His Grand Central plans dramatically improved passenger service to midtown Manhattan; now he proposed a small-car freight subway to solve Manhattan's freight transportation challenges.

Above all else, the plan treated freight transportation in the port of New York as a single system. No one railroad or shipping company could substantially reduce overall congestion. Each incremental change resulted in just a short-term gain. Only by treating the entire port as an integrated system could overall improvements be made. In order to succeed, the plan needed the cooperation of individual business interests and local municipalities, including the city of New York, even at the cost of sacrificing a short-term advantage.[20]

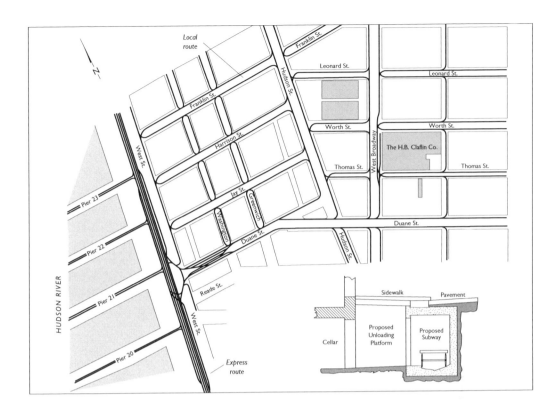

The small-car subway proposed by Wilgus for lower Manhattan, 1907: section running from Piers 20 to 23 on West Street, Duane to Franklin Streets, to West Broadway.

After the Amsterdam Company made the report public in September, Wilgus's plan received favorable press coverage. The *New York Times* devoted almost a full page to the proposal and included drawings of the freight subway under West Street with an elevated passenger railroad overhead. The *Times* described the proposal as a "gigantic plan to relieve street congestion" and asked why the problem of efficiently delivering freight in Manhattan had not been solved earlier, "this being the age of big undertakings successfully worked out."[21] The newspaper also described a new freight subway recently opened in Chicago that had been built by the telephone company as part of its franchise to locate telephone lines underground. This subway was much smaller in scale, with the capacity to carry only three tons of freight per car. Its tunnels ran almost 40 feet underground and required elevators to lift freight to all of the buildings served by the subway. By comparison, Wilgus planned to construct New York's small-car subway just below the sidewalks and even through store and factory basements. Each subway car, with a 10-ton capacity, could carry more than three times the capacity of the cars used in Chicago.

Engineering News, an influential professional journal, in an article detailing the small-car subway proposal in October, identified the central problem with all previous attempts to solve the west side problem. No proposed solution both "served the interest of traffic and preserved the rights of the railroad company."[22] Even after the passage of the Saxe Law gave the PSC the right to condemn the New York Central's tracks, the railroad vigorously defended its right to use the city's streets. If forced to eliminate trains on the streets, the railroad threatened to sue the city for damages, potentially a huge liability for New York.

An editorial in *Engineering News* pointed out that the Wilgus plan properly focused on the costs associated in moving freight to its final destination from the rail yards in the port of New York.[23] The editor, in a letter to Wilgus, identified another key issue: "solving the big problem of city transit and traffic." "This is," he continued, "the biggest problem that engineers have to solve at the present day." So successful had the railroad's efforts been to improve long-distance hauling that "it is not a great exaggeration to say that the cost of distance has been eliminated."[24] The real problem remained the handling of freight at the local level, the "last mile," precisely the problem the small-car subway eliminated.

Because of his high standing among his professional peers, Wilgus prevailed upon the editor of the *Railway Age Gazette* to publish his long article detailing the plan in October.[25] This article was a condensed version of the report prepared for the PSC.

Criticism of the plan soon arose as shipping companies, railroads, and freight distributers assessed whether they stood to gain or lose in the short run with the construction of the small-car freight subway. Soon after the plan became public, the truckers and dray operators met to voice their strong opposition to a plan that threatened to curtail their businesses.[26] The freight haulers who worked the piers and freight depots in lower Manhattan, ignoring the serious costs of congestion at the piers and on the city streets, vowed to block the construction of the freight subway.

The railroads, as short-sighted as ever, also raised serious objections. The president of the Delaware, Lackawanna & Western Railroad, W. H. Truesdale, wrote privately to Wilgus in July 1908, before the plan had been officially announced, and warned the plan would be opposed because "this is of course such a radical departure from present practice and involves such a large capital expenditure."[27] He argued that the cost estimate of $70 million would prove to be much too low, that the railroads would be reluctant to abandon their freight facilities in Manhattan, and that the proposed subway would not be able to serve all of their freight business. Wilgus wrote back to Truesdale and admitted his cost estimates to be "extremely rough."[28]

Since cost would be one of the determining factors if the small-car subway moved forward, Wilgus analyzed available data. First he needed to estimate the cost of the current freight delivery to Manhattan.

For all of the railroads serving the port of New York except for the New York Central, freight handling included lighterage charges to move the freight across the harbor from New Jersey to Manhattan. Once on the piers, individual shipments had to be unloaded from freight cars or barges, sorted, and loaded on drays for transport to final destinations. Shippers referred to this labor as the "break of bulk." The cost of lighterage and the "break of bulk" varied between $0.75 and $0.85 a ton. Wilgus used $0.80 a ton in his cost model. For carting freight from the piers on the city streets to final destination he used an average of $0.80 a ton (table 3.4).

Table 3.4. Wilgus's analysis of cost of distributing freight in Manhattan, 1908

	Est. cost per ton	No. of tons distributed (thousands)	Total cost (thousands)
Railroad merchandise			
Terminal costs on N.J. side of Hudson	$0.15		
Lighterage	0.80		
Terminal costs on waterfront Manhattan	0.50		
Cost to railroads	1.45		
Drayage (paid by shippers)	0.80		
Total cost, N.J. railroads and shippers	$2.25	9,800	$22,050
Coal			
Handling, storage, and lighterage	0.40		
Drayage	0.60		
Total for coal	$1.00	4,000	$4,000
Steamship merchandise			
Drayage	$0.80	9,400	$7,520
Grand total			$33,570

SOURCE: Wilgus, "Plan of Proposed New Railway System," 13.

The cost to the railroads and shipping companies to distribute 23 million tons of freight to Manhattan, within the port of New York, totaled over $33 million a year. In Wilgus's rational planning model, the railroads, steamship lines, and shippers would recognize the costs savings to be gained if the small-car subway system replaced the current cumbersome and expensive system of distributing freight. Thus, he expected most of the them to enthusiastically support the proposed freight subway.

To put these costs in perspective Wilgus pointed out that in 1907 it cost as much to move a ton of freight from a New Jersey rail yard to lower Manhattan as it did to haul a ton of freight between Boston and New York—a distance of 200 miles. The inefficiency of hauling freight in the port, "the last mile," added enormous costs.

The coal New York used for heating and lighting also incurred high handling costs on the waterfront. The cost of coal delivered to the waterfront, including profit to the mine owners and transportation, averaged $5.00 a ton. To handle the coal at the waterfront and then deliver it to consumers added $1.50 per ton, including a $0.50 a ton profit to the retailer. Wilgus argued that "the terminal costs alone, $1.00 per ton, is 20 per cent of the total price of coal delivered in the city." In his view, the high costs of handing freight in Manhattan and throughout the port of New York constituted a hidden tax. An improved distribution system could be paid for and a profit made from this hidden tax. Just as with using air rights—taking "wealth from air" at Grand Central—the small-car freight subway could be financed from the difference between the current high cost of moving freight to more reasonable charges for delivering freight by subway to the doorsteps of businesses throughout lower Manhattan.

Wilgus and the Amsterdam Corporation: From Consultant to Builder

As Wilgus continued to refine the plans for the small-car subway as part of the consulting contract with the PSC, both he and his partner in the Amsterdam Corporation, Henry Pierce, found themselves considering a completely different strategy. They believed that their plan offered the perfect solution to the west side problem. An enormous financial opportunity awaited the company which could secure the franchise, raise the capital, and build the freight subway. Why not do it themselves? Instead of just serving as consultants and then watching others make millions, they decided to pursue the golden opportunity themselves.

In early June, Wilgus and Pierce visited J. P. Morgan in his office on Wall Street to discuss their plans for the freight subway and to solicit Morgan's financial support. If they could gain his support, their plan stood a good chance of moving forward. On June 23, two months before their report was due to the PSC, Wilgus wrote a confidential letter to W. H. Newman, president of the New York Central and Wilgus's former boss, stating the intention of the Amsterdam Company to pursue a franchise to build the freight subway and raise the necessary capital. He suggested that the Central and the other railroads in the port seriously consider investing in the subway but asked Newman to keep all of this quiet for the time being. Wilgus proposed that the railroads, including the New York Central, consider investing at least $50 million in the project "to insure their control." Given that the Central continued to face the prospect of losing the right to run its trains on city streets to the St. John's Park depot, Wilgus assumed that "the New York Central should be especially interested in the project because if successfully carried out, it would obliterate a profitless expenditure by the NYC of from $10,000,000–$20,000,000."[29] The expenditure referred to negotiations between the railroad, the PSC, and the city to have the Central build a train subway between their 60th and 30th Street freight yards.

Just two days after sending the letter to Newman, the Amsterdam Corporation billed the PSC $3,000 for consulting work completed to date.

Before Wilgus and Pierce could proceed with their proposal to build the freight subway, they had to resolve the problem that stood in their way: the Amsterdam Corporation had a consulting contract with the PSC, the very state commission that controlled the award of all transportation franchises in New York City. How could they argue the merits of the small-car freight subway in their consulting-contract report and then turn around and submit a petition to the very same commission for a franchise to build and operate the freight subway? An obvious conflict of interest needed to be resolved. Wilgus did not care about the conflict of interest; he saw only the opportunity to make millions. A small consulting project for the commission paled by comparison. Pierce agreed: "I agree with you that the matter is of such immense importance to us that we must not and will not be sidetracked by the Commission."[30]

Wilgus finally decided to take the direct approach and wrote to Edward Bassett, chair of the commission, on September 10, 1908, to inform him of the plans of the Amsterdam Corporation. Never modest about his own work he explained to Bassett: "I am so impressed with its practicability [that of the small-car freight subway] that I am desirous of associating myself with others to a view to submitting a proposition to the Commission after the report has been made public."[31] Wilgus offered to return the $3,000 consulting fee and asked Bassett to return the draft report. Bassett and the PSC agreed to accept the refund and cancel the consulting project, but added a note of rebuke: "We regret that you could not have told us at the beginning that there was a possibility of you connecting yourself with a business enterprise."[32]

Wilgus and Pierce made their proposal public at the beginning of October and immediately received favorable press coverage and a number of solicitations from potential investors. The president of the Pennsylvania Railroad, the largest railroad in the country, with a significant investment in freight facilities in the port, visited Wilgus in his office to go over the proposal and cost estimates in detail. E. V. W Rosseter, vice president of finance of the New York Central, wrote a letter of support and suggested that the Central might invest in the project: "If you can get the right kinds of privileges and franchises from the authorities there may be no trouble about the money end."[33]

Rosseter identified the key hurdle Wilgus and Pierce faced. Building a freight subway under the city streets and sidewalks, a tunnel under the Hudson River, and a belt railway in New Jersey required the formal approval of the federal government, two state governments, New York State's Public Service Commission, and the mayor and the Board of Estimate in New York City. They also needed cooperation and support from the ten railroad companies and hundreds of steamship companies and shippers in the port. In addition, powerful political forces needed to be convinced of the merits of the plan, not to mention the general public, who, in New York, were never hesitant to voice their opinions. The latter constituencies would be difficult enough, but the formal approval required from the PSC and the Board of Estimate constituted an enormous hurdle; unless they awarded a franchise, the plan would be dead. Rosseter voiced his pessimism about winning approval: "In these days of jealousies on the part of Municipal authorities and their reluctance to grant any privileges from which there would seem to be a prospect of profit in the far future, it is awfully hard to get permission to carry out any scheme like this."

Dealing with the local political forces in New York proved to be a daunting challenge; the question of private versus public enterprise could never be settled. Building a transportation infrastructure in New York City or anywhere in the country demanded capital on an unprecedented scale. Once a rail system was constructed, operating it required skilled management and constant investment for maintenance and improvements. Should railroads, elevated railroads, or subways be financed by government and owned and operated to serve the "public interest?" Suspicion of the government, especially local government in cities like New York, ran deep in the nineteenth and early twentieth centuries, the very time of dramatic population growth and industrialization. New York's history details the constant effort to construct an urban infrastructure to keep pace with the city's growth, with politics never far from the center of the process. On the other hand, large corporations, the railroads above all, also came under intense public scrutiny. If private companies were to own mass transportation, critics would argue that they profited unfairly at the expense of the public.

The city and the state of New York had never settled on one transportation model. When the Erie Canal was constructed, DeWitt Clinton and other political leaders knew private enterprise could not raise the necessary capital; the costs were simply too daunting. The only way to build a four-hundred-mile canal across upstate New York in the 1820s was to have the state borrow the necessary funds, contract for the construction, and then run the canal as a state enterprise. To be sure, the Erie proved to be a financial success and generated a profit from the very beginning, ensuring the state of New York with sufficient funds to pay off the bonds sold to finance the canal. Almost as soon as the Erie Canal opened, political pressure came from other parts of the state for additional canals to be built. New York responded to the pressure and constructed hundreds of miles of canals, none of which proved to be financially sound and were kept operating only by using the surplus generated by the Erie Canal. Even when a local canal lost substantial amounts of money each year, it proved very difficult to shut down. Local politicians put pressure on the legislature to continue subsidies no matter how great the losses.

When the railroad revolution followed in New York, private enterprise replaced government, but with important caveats. First, a railroad needed a state charter to form a corporation and raise capital to build. The railroad also needed to acquire private property for track rights over long distances at a reasonable cost. When railroad companies encountered private property owners unwilling to sell or demanding an exorbitant price, the state often intervened and used its power of eminent domain to enable the railroad to acquire the needed property. Within cities railroads also needed to acquire private property and in many cases to use public property. The city and state charters for both the Hudson River Railroad and the New York & Harlem Railroad gave them the right to build their tracks on the streets of New York. Whenever a transportation company needed to use public space, the municipal government had to acquiesce and grant a "franchise"—permission to utilize public space for private enterprise.

Despite the awarding of franchises, the railroads often encountered difficulty raising needed funds and pressured the state or local governments to provide guarantees to private investors or to invest directly in the transportation project. In the case of New York City's first subway, the city invested almost $37 million to build the subway. August Belmont's IRT purchased the necessary equipment and operated the subway under a long-term contract with the city—a public-private partnership that ensured controversy.[34] Even with the universal acclaim and praise the IRT garnered when first opened in 1904, suspicion soon followed.

Why should the IRT "profit" at public expense? No matter what the fare to ride the subway, politicians and the public viewed the IRT as a "monopoly" that extracted unfair tribute from ordinary citizens.

Franchise Application to the Public Service Commission

As soon as the PSC agreed to cancel the consulting contract with the Amsterdam Corporation, Wilgus and Pierce informed the commission that they planned to formally seek approval for their small-car subway system. The granting of a franchise by the commission was the absolutely crucial first step in their efforts to build the freight subway as a private enterprise.

Wilgus carefully revised his cost estimates and arrived at a figure totaling over $100 million, a staggering sum (table 3.5). (By comparison, in 1903 Wilgus had estimated the total cost of the Grand Central project at $34 million, and as noted above, New York City had invested $37 million to build the IRT.) Building the freight subway would be an undertaking as complicated and expensive as the entire Grand Central project, including the construction of the electric zone in the Bronx and Westchester County and the building of Terminal City. Wilgus added a caveat recognizing the ever-escalating cost of doing business in New York: "Inasmuch as general experience with large enterprises in New York has shown a tendency for work to cost more than first estimated, it has seemed wise to arbitrarily add a further contingency fund of $20,000,000."[35] Given his experience at Grand Central—where his initial estimate of $34 million had increased to over $70 in 1907 just before he left the railroad—adding this contingency seemed fiscally prudent.

In his franchise application Wilgus listed the advantages of the small-car freight subway:[36]

Present Method (Waterfront delivery to drays)	*Proposed Method* (Subway delivery to consignee)
a. Received in yard at "Meadows or N.J. waterfront	a. Received in yard at "Meadows" [Hackensack Meadows]
b. Classified in yard at N.J. waterfront	b. Classified at yard in "Meadows"
c. Transferred to float at N.J. waterfront	c. Placed at transfer facility at "Meadows"
d. Floated across North River [Hudson River]	d. Trans-shipped from large to small car at "Meadows"
e. Trans-shipped, car to pier in Manhattan	e. Subway car sorted in classification yard
f. Trans-shipped, pier to dray in Manhattan	f. Hauled via subway to consignee
g. Drayed, pier to consignee	

As Wilgus planned, switching freight in the New Jersey train yard eliminated the need for the railroads to operate their expensive system of lighters, barges,

Table 3.5. Amsterdam Corporation cost estimates for construction of small-car freight subway (thousands of dollars)

	Specific costs	Totals
Hackensack Meadows		
Right-of-way	$1,600	
Grading, masonry	2,750	
Tracks, signals, interlocking	1,000	
Interchange yards, "Subway"	1,320	
Classification yard, shops	1,430	$8,100
Bergen Hill Tunnel		
Tunnel proper, easements	1,600	1,600
Connection to Hoboken waterfront		1,000
North River Tunnel, Hudson River		4,800
60th Street to Battery, Hudson River		
4-track subway	7,500	
Connection to piers, junctions	4,000	11,500
Battery to 33rd St., East River		
2-track subway	3,500	
Connection to piers	1,500	5,000
Cross-town line, 2-track subway		1,500
Bronx Branch, 2-track subway, 1st Ave.		4,600
Distributing system beneath sidewalks in lower		
Manhattan (30 miles of streets, 60 miles of track)		16,500
Readjustments of surface structures		4,000
Electrification, exclusive of generating		4,000
Equipment: 3,000 motor cars, 1,500 trailer cars,		
1,200 autotrucks		13,500
32 transfer stations		6,000
Interest on money during construction		8,900
Contingencies		9,000
Grand total		100,000

SOURCE: "Brief Submitted on Behalf of the Amsterdam Corporation by William J. Wilgus, to the Public Service Commission for the First District," Dec. 26, 1908, 17–18, William J. Wilgus Papers, box 45.

and tugboats. With almost all freight hauled under the Hudson River rather than floated across, the railroads no longer needed the small armada of vessels currently crowding the waterways of the port. On Manhattan Island and along the Brooklyn waterfront, shipping companies could expand and take over valuable pier space occupied by the railroads. The additional piers would ensure the port of New York's dominance of shipping in the United States.

Wilgus's plan to finance the small-car subway assumed that a major share of all freight distributed to the port of New York would switch from the present distribution system to the new subway. The cost per ton for using the freight subway would be lower than the current costs paid by the railroad. Even with lower rates, Wilgus's cost model projected revenue sufficient to pay interest charges on the borrowing used to finance the project, to cover all operating costs, and to produce a surplus (table 3.6). Construction of such a complicated project would take years; Wilgus projected a completion date of 1912. He adjusted his estimates of freight tonnage from 1907 data to 1912 assuming a 4 percent increase each year (table 3.7). The total estimate of 24,650,000 tons of freight to be delivered by the new freight subway assumed 90 percent of all of the freight handled by the ten railroads on the New Jersey side of the harbor switching to the subway. If even two or three of the railroads refused to cooperate or did not see the advantage of the new subway service, the estimated freight volume would decrease significantly threatening the financial plan.

Table 3.6. Wilgus's cost model for small-car freight subway

Receipts

8,000,000 tons of merchandise to & from N.J. railroads via subway	$1.40/ton	$11,200,000	
2,600,000 tons of merchandise to & from N.J. railroads via subway & autotruck	$1.50/ton	$3,900,000	
6,000,000 tons of merchandise to and from water carriers via subway	$0.60/ton	$3,600,000	
2,000,000 tons of merchandise to and from water carriers via subway & autotruck	$0.70/ton	$1,400,000	
3,800,000 tons of coal	$0.50/ton	$1,900,000	
2,000,000 tons of waste material	$0.50/ton	$1,000,000	
250,000 tons of mail and express	$2.00/ton	$500,000	
A total of 26,500,000 tons			$23,500,000

Expenditures

Bonded interest, 5% on $120,000,000		$6,000,000	
Operating Costs			
40,000,000 car miles at $0.10	$4,000,000		
6,500,000 tons by autotruck	$2,600,000		
5% depreciation on $30,000,000	$2,000,000		
Taxes	$2,000,000		
General expenses	$1,400,000	$12,000,000	$18,000,000

Surplus | | | | $5,500,000 |

SOURCE: "Brief Submitted on Behalf of the Amsterdam Corporation," 31.

Table 3.7. Wilgus's estimate of tonnage delivered to Manhattan via small-car subway in projected first year of operation (tonnages in thousands)

Origin	1907 tonnage	Total 1912 tonnage	1912 deliveries by subway			
			Subway	Subway & truck	Total	% of 1912 total
Railroads						
Merchandise	9,800	11,900	8,000	2,600	10,600	90
Coal	4,000	4,800	1,900	1,900	3,800	80
Total	13,000	16,600	9,900	4,500	14,400	
Water carriers						
Transatlantic	2,200					
Coastwise	2,900					
LI Sound	1,400					
Hudson River	1,000					
Other	6,400					
Total	13,900					
Transferred to & from railroads	12,400	14,900	5,500	2,000	7,500	50
Sub distribution by drays	800	1,000	500		500	50
Waste	4,000	4,000	2,000		2,000	50
Mail and express	300	500	250		250	50
	31,300	37,000	18,150	6,500	24,650	66

SOURCE: "Brief Submitted on Behalf of the Amsterdam Corporation," 30.

Not only would the small-car freight subway turn a substantial profit each year, but other benefits would also follow. Consolidating freight in one rail yard would increase the use of each railroad's rolling stock. Under the current system rail cars often sat idle in the rail yards in New Jersey waiting to be moved to the shorefront to be floated across the harbor to Manhattan. Railroads interchanged freight cars and then paid a per diem if another railroad's car stood idle in a rail yard. Delays moving freight cars through the railroad system increased per diem costs. Empty cars needed to be recirculated as quickly as possible; empty freight cars produced no revenue. In the port of New York, the thousands of empty cars that filled the railroad's yards caused a ripple of delays. At times, shippers in Chicago could not move their merchandise because the railroads' freight cars were standing idle in New York.

The New York Central stood to gain from the freight subway as much as any of the other railroads. A freight subway down the west side of Manhattan, from the railroad's 60th Street yard, would eliminate the need to use the city streets to haul freight to the lower end of the island. Wilgus believed the subway solved the company's long-running west side problem and expected the Central to support the small-car subway plan.

Businesses in Manhattan would also gain from the removal of freight delivery from the congested city streets to an underground delivery system. A much more efficient delivery system meant goods would arrive in a timely manner, increasing efficiency. While a shipping company could calculate the costs of delays from wage records and maintenance costs, estimating the gain from moving freight more quickly around the city was difficult. Nevertheless, Wilgus projected that annual savings for consumers would be at least $5.2 million. This figure represented the difference between freight subway charges per ton and the current cost of moving a ton of goods by dray.

And it was not just that moving tonnage by dray was more expensive. The thousands of drays that traveled the streets of lower Manhattan contributed to the overall congestion of the great city and resulted in tons of horse manure in city gutters. With numbers of drays removed, the quality of life for the general public would improve. Even though any "generalized good" could not be easily quantified, less crowded and less polluted streets provided benefits for all citizens. Wilgus argued that with less congested streets, property values would increase. With railroad freight handling removed from the piers along the waterfront, shipping companies would have space to expand, reversing the decline in the port of New York's share of foreign and coastal trade.

The brief that the Amsterdam Corporation prepared for the PSC estimated the total savings to be gained from a small-car freight subway at over $11 million. Add the projected surplus of $6 million, and Wilgus estimated total "annual monetary savings and surplus" at $16.4 million. While this figure may seem to be fanciful and included indirect savings rather than hard dollars and cents, Wilgus continued to present the small-car subway as a radical, innovative plan to completely transform the movement of freight in the entire port of New York and in lower Manhattan, the core of the region's business.

Without any delay, the PSC scheduled a formal hearing to consider the proposal of the Amsterdam Corporation for late December 1908. Meanwhile Wilgus and Pierce continued to meet with executives of the individual railroads in efforts to build support for the plan. Wilgus published a major article in *Railway Age Gazette* to inform his professional peers about the proposal.[37] They also met with potential investors. Even if the PSC approved their request for a franchise, the daunting task of raising more than $100 million remained.

Politics and the Failure of the Wilgus Plan

No history of New York in the time period from the end of the Civil War to the turn of the twentieth century can avoid the subject of municipal corruption. Boss Tweed, the "Forty Thieves," Tammany Hall, the Tweed Courthouse—all evoke the taint of corruption on a monumental scale, in both the city and the state of New York. For a period of time, no major construction project in New York moved forward without tribute paid to the reigning political machine. Commodore Vanderbilt carried carpet bags full of money to Albany in 1869 to grease the merger of his Hudson River and New York Central railroads. All major transportation projects needed the approval of city and state government to proceed. Building a mass transit infrastructure in New York involved overcoming daunting engineering challenges and at the same time remained enmeshed in the gritty political life of the city of New York.

The New York Central Railroad needed to secure multiple approvals from the city of New York before the Grand Central project started. In his memoirs Wilgus gives no hint of any corruption involved while he worked for the Central planning and then directing the massive construction at 42nd Street. A far different experience awaited him in dealing with the PSC.

The commission had asked Wilgus to testify at the December public hearing that was to consider the Amsterdam Corporation's application for a franchise to build the small-car subway. As Wilgus worked on a brief to be filed in support of their application and Pierce continued to meet with potential investors, their entire effort abruptly came to a halt. Political corruption appeared in the person of the secretary to W. R. Willcox, the chairman of the PSC. As Pierce waited in the office for a meeting with Willcox, Mr. Farnum, the secretary of the PSC, spoke to him and strongly suggested that Pierce meet with a Richard Wood, a stockbroker, "who would be of great assistance to the project."[38]

Wilgus, a number of years after the incident, wrote a memo for his files detailing the bribery attempt. Pierce reported a visit to the offices of the Amsterdam Corporation by Wood, who claimed he could be of "great service" with the PSC in exchange for a significant amount of stock in the Amsterdam Corporation, which would be very valuable once the company secured a franchise for the small-car freight subway. To prove his value, Wood brought Pierce and Wilgus a copy of the confidential report prepared by the commission's chief engineer, evaluating the small-car freight subway proposal. The report included the chief engineer's recommendation that the plan be approved by the PSC and that Wilgus and Pierce be awarded the franchise to build the small-car freight subway.

Wilgus and Pierce clearly knew exactly what Wood wanted. If they did not agree to pay Wood, he would work behind the scenes to block the PSC from approving their application. Just as Commodore Vanderbilt's money furthered his railroad interest with the politicians in Albany, it appeared to Wilgus and Pierce that the tentacles of political corruption reached to the office of the chairman of the PSC.

Without any hesitation Wilgus and Pierce decided not to agree to the demands made by Wood, no matter the consequences. In his memorandum, Wilgus summarized their decision: "Mr. Pierce and I agreed that under no circumstance would we be party to any such arrangement as that proposed by Mr. Wood." He added one more sentence: "No action thereafter was taken by the Commission on the company's application."[39] Refusing to agree to Wood's proposal meant that their plan for a small-car freight subway to solve the west side problem would be rejected. While they refused to bow to corruption, a real opportunity to fundamentally change transportation in the port of New York passed by.

A number of years later, in 1914, Governor Glynn appointed Richard C. Wood to the PSC.

The West Side Problem—"To Be Continued"

The demise of the Wilgus and Pierce plan did not end efforts to solve the west side problem; congestion along the Hudson River continued to increase year after year. In 1910 Calvin Tomkins, commissioner of docks for the city of New York, proposed building a four-track elevated railway along the river and five terminal buildings, connected to the elevated railroad, between 25th and 30th Streets along 10th Avenue. Tomkins's plan assumed that all the New Jersey railroads would share the terminal buildings and that they would float their freight across the Hudson River to railroad piers between 37th and 40th Streets.

Critics of the Tomkins plan, Wilgus included, pointed out that freight would still have to be distributed by drays on the streets of lower Manhattan from the freight terminals. The Tomkins plan closed West Street to all vehicular traffic between 25th and 30th Streets opposite the freight buildings to allow freight cars to be moved from the freight piers onto the elevated tracks. Wilgus pointed out how the north-south avenues contributed to efficient transportation: "The configuration of Manhattan Island and the location of its streets are such as to make imperative less rather than added burdens on the north-south arteries of the city, which are comparatively few in number."[40]

In the early 1900s, New York City expanded the width of West Street along the Hudson to improve access to the piers and to improve the flow of traffic up and down the island. Given the dense development along the interior avenues in Manhattan, no possibility existed for improving the flow of north-south traffic by widening the major avenues. Only along the Hudson and East River waterfronts could subways, elevated railroads, and eventually highways be built to speed passengers and freight up and down the island.

Negotiations dragged on between the city of New York and the New York Central Railroad over the west side problem. The city continued to demand that the railroad remove its tracks from the city streets. In September 1911, the railroad presented a plan to build an elevated railway along the Hudson River and roof over the company's tracks north of 72nd Street, extending Riverside Park to the edge of the river. South of the rail yard at 30th Street a two-track elevated rail line would be built along 9th Avenue using track rights over private property. Negotiations continued for years until World War I intervened.

After World War I, efforts to find a solution to the west side problem resumed, and in 1917 the New York State Legislature appointed a commission headed by William H. Benschoten to undertake yet another investigation of the freight situation in Manhattan. The commission's detailed report recognized, just as Wilgus had in his 1907 proposal, that any solution to the freight congestion had to include the railroads in New Jersey. One key recommendation involved establishing a terminal commission with authority from the state of New York to build a "comprehensive and modern freight system, not only for the New York Central lines, but also for the railroads which float their freight across the Hudson River."[41] The report stated forcefully that if the railroads serving the port of New York did not see the advantage of using the new system, the states and federal government had the power to force them to use the joint facilities.

In 1907 Wilgus had assumed that the railroads would voluntarily use his small-car freight subway because his careful analysis logically demonstrated the financial and operating advantages to be gained. The 1917 commission did not rely on logic and analytical persuasion when it came to overcoming the narrow, parochial interests of the railroads. They recommended that a terminal commission be delegated the power to force the railroads to cooperate. Wilgus had believed that the small-car freight subway would attract private investment and would be operated as a private enterprise. Just over a decade later, proposals to solve the transportation needs in the port of New York shifted decisively. If the private railroad companies and shippers using the port refused to cooperate, a "commission," a quasi-governmental agency, would be delegated the power to improve transportation and, if need be, force the railroads and shippers throughout the port to cooperate. By definition, a commission consisted of non-elected officials appointed by the governor with the necessary power to implement solutions. The future of the port of New York would no longer be directed by private enterprise or local elected officials; rather, an all-powerful commission, officially dedicated to a rational planning model, would assume responsibility. In 1919, the states of New York and New Jersey took the fateful step and created the New York, New Jersey Port and Harbor Development Commission to conduct a detailed study of the transportation needs of the entire port and recommend steps for improvement, including the establishment of a permanent port authority.

The High Line and Highways

Even with the creation of the Port of New York Authority in 1921, freight trains remained on the streets of the west side of Manhattan. "Death Alley" maintained its reputation for mayhem as the New York Central insisted on its right to haul freight to and from lower Manhattan on the city streets. In 1919 the recently established New York, New Jersey Port and Harbor Development Commission, the immediate precursor to the Port of New York Authority, provided detailed information on freight handled on the island below 60th Street. The New York Central had moved 1,672,445 tons of freight in 270,988 freight cars, hauled north and south on the city streets, in just one year.[42] Over the next decade, despite the work of the Development Commission and the establishment of the Port Authority, the city and the New York Central could not reach agreement on any solution to the west side problem.

Finally in 1929 after endless bitter debate and the threat of lawsuits, the city, the state, and the New York Central Railroad agreed on a massive West Side Improvement Project. A key part of the plan involved building the "High Line," an electrified, two-track elevated railway from the Central's 30th Street rail yard to a new St. John's Park freight depot just north of the Battery. Rather than build the High Line over city streets, the city and railroad negotiated with property owners to construct the elevated line in the middle of the block, in some cases right through factories and warehouses. The improvement included a subway tunnel connecting the Central's 30th and 60th Street rail yards and roofing over the railroad tracks in Riverside Park, adding 23 acres of recreational space.[43] Removing the Central's tracks from the city's streets eliminated 105 street-level railroad crossings and ended the danger on "Death Alley." Building the High Line proved to be a massive undertaking that involved over five years of construction and costs estimated at $150 million.

In his memoirs Wilgus dryly pointed out that the cost of the High Line exceeded his estimated cost for the small-car subway and provided only a partial solution to the problem of hauling freight to Manhattan. More importantly the costly High Line proved not to be an integrated solution to handling freight in the port. The High Line was used exclusively by New York Central freight trains. The railroads on the New Jersey side of the harbor could not use these tracks and continued to float their freight across the harbor, monopolizing valuable pier space for freight handling. In addition, the railroads had to continue to absorb all of the costs of handling freight in order to provide freight service to Manhattan and Brooklyn.

The West Side Improvement Project agreement in 1929 came just two years after the opening of the Holland Tunnel in 1927, linking the New Jersey shore of the Hudson River to Manhattan Island at Canal Street. The Holland Tunnel represented revolutionary change for the port of New York's transportation system: the tunnel was a vehicular tunnel and not a railroad tunnel. In a flash, freight could be trucked under the Hudson instead of being floated across the river. In the same fateful year of 1929 another innovative transportation project—the West Side Highway—eventually led to the downfall of the High Line. Along the Hudson River where Wilgus had planned his small-car freight subway, the West Side Highway and Holland Tunnel provided access for a growing fleet of trucks, which soon dominated the port's freight hauling. The High Line proved to be an albatross.

Massive change came to freight and passenger transportation in the form of the truck and car. Public spending facilitated the rapid change by shifting investment from mass transportation to roads, highways, bridges, and tunnels. The railroads, still private companies, simply could not compete with this massive commitment of public finance to the truck and automobile. Although Wilgus continued to advocate for railroad solutions to New York's transportation challenges, he fought an uphill battle. Even an engineer who had built his career in railroading and had planned the Grand Central project as a solution to New York's passenger rail service would be drawn to the automobile age. In the 1920s, Wilgus served on the Board of Consulting Engineers for the Holland Tunnel. His engineering talents and his analytical approach proved as valuable to the building of the longest vehicular tunnel in the world as they had to the construction of the largest railroad terminal in the world.

4

EXPANDING THE SUBWAY IN MANHATTAN

NEW YORK'S transportation challenges extended beyond the need for roads and rail lines tying the city with the rest of the country and with the city's growing suburban areas on Long Island, in Westchester County, and across the Hudson in northern New Jersey. Over 500,000 people lived in Manhattan in 1850; by 1880 the population had doubled to 1,164,673, dramatically increasing the need for better transportation on Manhattan Island.

Over a number of decades, different modes of transportation evolved to move passengers more efficiently on the city streets, above the streets, and eventually below the streets. From the humble omnibus in 1827 to the opening of the city's first subway, the Interborough Rapid Transit, in 1904, public and private investors created a complex transit system with a history riddled with layers of political intrigue and corporate machinations.

When the IRT began service, the subway seemed to offer truly rapid transit and to be the ideal solution to moving passengers around the most crowded island in the world. The story of the New York City subway system involves a convoluted tale. Clifton Hood's masterful history of the city's subways, *722 Miles*, details the difficulty of financing, building, and operating what eventually came to be the largest subway system in the world.[1] The first subway was an immediate success, with hundreds of thousands of riders each day, and led to demands for an expanded system, not only in Manhattan, but out to the other boroughs. For the next nine years a battle royal ensued over how and where and by whom to expand the subway.

Never hesitant, in 1909 Wilgus stepped into the subway controversy and proposed building a combined passenger and freight subway and an elevated railroad to encircle lower Manhattan. With the Amsterdam Corporation's proposal for a small-car freight subway still officially before the Public Service Commission, Wilgus presented the commission with another bold plan to expand subway service and at the same time solve the city's west side freight problem.

Wilgus argued: "No system appears to have been devised for efficiently caring for internal circulation on the Island, between various rail and water terminals now established or proposed in connection with future subways and bridges."[2]

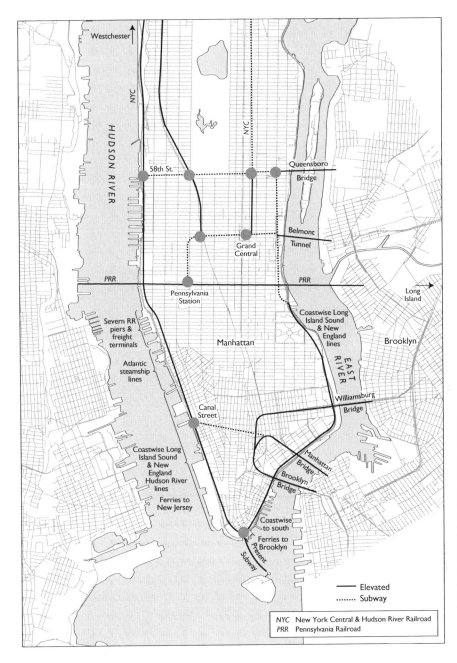

Westchester

HUDSON RIVER

NYC

NYC

58th St.

Queensboro
Bridge

Belmont
Tunnel

Grand
Central

PRR

Pennsylvania
Station

PRR

Long
Island

Severn RR
piers &
freight
terminals

Coastwise Long
Island Sound
& New
England
lines

Manhattan

Brooklyn

Atlantic
steamship
lines

EAST RIVER

Williamsburg
Bridge

Canal
Street

Coastwise Long
Island Sound
& New
England
Hudson River
lines

Manhattan
Bridge

Brooklyn
Bridge

Ferries to
New Jersey

Coastwise
to south

Ferries to
Brooklyn

Present

Subway

——— Elevated
.... Subway

NYC New York Central & Hudson River Railroad
PRR Pennsylvania Railroad

Wilgus's proposed Inter-Terminal Belt Line, 1909: connecting piers along the Hudson River south of 58th Street, piers and bridges along the East River, and subway connections to Grand Central Terminal and Pennsylvania Station.

His proposed "Inter-Terminal Belt Line" would connect all major terminals, bridges, and ferries to move people around the city much more efficiently, especially in lower Manhattan. Wilgus dismissed previous efforts to improve passenger circulation in Manhattan. For well over half a century a series of new transportation technologies had improved transportation on the island, but none had provided a perfect solution to "internal circulation."

Improving Internal Circulation on Manhattan Island

Mass transportation in New York City began in 1827, when horse-drawn wagons with benches for passengers—omnibuses—appeared on Broadway, carrying passengers between Battery Park and Canal Street. The omnibus provided transportation along a fixed route for a standard fare and facilitated the settlement of areas of the city to the north of lower Manhattan. Soon, omnibuses were available on all of the city's north-south avenues and on many cross streets. While an improvement over walking, the omnibuses added to the overall congestion on the city streets, competing for space with horse-drawn drays carrying freight, private coaches, and hackney cabs for hire. As the *Times* reported, "So numerous have [omnibuses] become, that the great thoroughfares of this City are now mostly occupied by them."[3]

On Broadway, traffic congestion created dangers for both vehicles and pedestrians: "Each month and each week beholds the accession of new crowds and businesses to those which already choke Broadway to suffocation. . . . You were more jostled today than yesterday. To cross Broadway, in any of its lower sections, is a task of infinite hazard. . . . Sometimes you spend half an hour on the sidewalk, waiting a favorable opportunity. . . . Vehicles of all denominations are inextricably interlocked. Forward or backward no progress can be made."[4] This passage could have been written in 2011, but it appeared in April 1852, describing New York's crowded streets, especially the city's famous Broadway. The editorial lamented the dust on the streets: "The thick smudge penetrates everywhere." Omnibuses driven recklessly threatened life and limb, as did "a butcher's cart, in rapid motion, driven by some reckless boy." Another editorial in 1860 noted that traveling downtown by omnibus "can hardly be less than three-quarters of an hour. . . . The entire upper part of the island offers every advantage to residents, except accessibility."[5] New York desperately needed additional forms of "rapid transit" to improve travel throughout the city.

Street Railways

Beginning in the 1830s one major improvement did provide faster transportation: the street railway. The city of New York awarded the first franchises to construct street railways in 1831. Each franchise granted a private company a right-of-way to lay rail tracks on the city's streets for horses to pull railway cars. With dedicated space on the street, the street railway theoretically promised faster service than the omnibus.

In April of 1831, the New York State Legislature passed a bill to charter the New York & Harlem Railroad Company with the right to build a street railway from 23rd Street to the Harlem River between 3rd and 8th Avenues.[6] The charter allowed the new railroad to use either horse or steam power. A year later, the legislature amended the original charter to extend the line south to 14th Street and even further south, if the city approved. The charter required the company to use horse power and not steam south of 14th Street, with a speed limit of five miles an hour.

The Harlem decided to lay its tracks on 4th Avenue from 23rd Street to the Harlem River. Of course, 4th Avenue later became the famous Park Avenue. From this humble beginning, the New York & Harlem Railroad eventually became part of the Vanderbilt railroad empire. On land purchased at 42nd Street by the New York & Harlem, Commodore Vanderbilt opened Grand Central Depot in 1871 and later William Wilgus planned the new Grand Central Terminal.

On November 26, 1832, the Harlem Street railway ran its first cars from Prince Street in lower Manhattan to 14th Street with the mayor and other dignitaries aboard. Two years later, the Harlem line reached Murray Hill, just a few blocks south of 42nd Street. In 1837 the Harlem's tracks arrived in the village of Harlem at the northeastern end of Manhattan Island. Eventually, the railroad crossed over the Harlem River into the Bronx, and then up into Westchester County to White Plains in 1844. What started as a small street railway in lower Manhattan became a railroad connecting New York to Westchester County and then farther north into Putnam and Dutchess Counties.

The success of the Harlem encouraged the omnibus companies to convert their service to street railways: on 6th Avenue in 1851, 2nd and 3rd Avenues in 1853, 8th Avenue in 1855, 9th Avenue in 1859, and 7th Avenue in 1864. Service on the major cross streets followed, and soon "lower Manhattan was a gridiron of street railways."[7] Eventually, cables replaced horses as motive power. Then in

Building cable-powered street railways on Broadway, 1891. Workers are digging a trench for the cable gripped by the street railway cars; this image looks south at 18th Street toward Union Square.

Table 4.1. Manhattan street railways, 1857

Railway	Miles of track	Cars	Horses	Passengers	Car miles	Receipts	Net
8th Ave.	4.89	61	471	6,879,452	1,023,157	$341,471	$147,088
6th Ave.	4.00	98	354	5,250,278	950,000	262,041	99,981
3rd Ave.	6.00	17	529	8,155,515	1,570,000	405,378	138,184
2nd Ave.	9.00	58	368	3,367,371	985,500	168,368	58,368
Total	23.89	234	1722	23,652,616	4,528,657	$1,177,258	$443,621

SOURCE: "City Railways," *New York Times Supplement*, Apr. 16, 1859, 2.

1886, the first electric-powered street railcars appeared, soon to be referred to as streetcars or trolleys.

Street railways were a decided improvement over the omnibus. Although traffic continued to be congested on the city streets, the street railways offered some semblance of faster service. The street railways also operated multiple cars during rush hours in Manhattan. Soon street railway companies began service in the other boroughs. In Brooklyn a number of companies built tracks on the major streets connecting to all of the ferry terminals to Manhattan. The Manhattan-based companies transported millions of passengers each year and were profitable enterprises. An 1859 report for the 2nd, 3rd, 6th and 8th Avenue Railways, printed in the *New York Times*, illustrates the scale of the enterprises (table 4.1). In this one year, the four street railway companies carried over 23.5 million passengers and grossed over $1 million. With only 24 miles of track, the streetcars carried 985,525 passengers per mile. No other rapid transit system in the world approached a similar volume. Critics valued the New York street railway franchises at over $100,000 per mile and pointed out that the fares paid by the working people of the city, at five cents a ride, provided a return of 10 or 15 percent to the owners.

While street railways provided improvements over the omnibus, they were equally subject to the congestion on the city streets and often faced delays when other vehicles or pedestrians blocked the tracks. For truly rapid transit, rail lines had to be separated from the flow of traffic on the city streets. Tracks could either be placed underground in a tunnel, as in London in 1863, or elevated over the streets.

Elevated Railways

The relentless demands for improved transportation on Manhattan Island led to the building of the city's first elevated railway on 6th Avenue.[8] The company that owned this line was chartered as the West Side and Yonkers Patent Railway in 1867 (it later changed its name to the New York Elevated Railway). On September 6, 1869, the elevated conducted a test run on 6th Avenue. Invited guests and the directors found the ride, from the Battery to Cortland Street, to be "remarkably smooth" on the single track in operation.[9] Commentators pointed out that the new elevated rail cars tied together a number of the city's major business and commercial districts. At 6th Avenue and 14th Street, the company had "erected one of its largest and most tasteful stations and close at hand is the extensive establishment of R.H. Macy & Co."[10]

The New York State charter granted the company the right to charge a fare of five cents for two-mile rides and not more than ten cents for all other distances on Manhattan. In return, the company paid 5 percent of its net income to the city of New York for the right to use the "air" above the city streets.

New Yorkers rarely complained about climbing stairs to the stations, 30 feet above the streets, because the "Els" provided the first true rapid transit on the island. Initially, cables pulled the elevated rail cars, but soon steam engines replaced the cables and obtained speeds of 20 miles an hour, moving passengers much more rapidly to and from lower Manhattan.

Other elevated lines followed the one on 6th Avenue: on 7th and 3rd and eventually on 9th and 2nd Avenues. By 1871, the Manhattan Railway Company consolidated control of all of the elevated railways. Cyrus Field, renowned as a founder of the Atlantic Telegraph Company and for laying the first successful transatlantic telegraph cable in 1858, became president of the Manhattan Railway Company in 1877.

A franchise to operate mass transit in New York City proved to very valuable. All early rapid transit companies—street railways, elevated railroads, and steam railroads—used the "operating ratio," the percentage of expenses to revenue, as a key indicator of financial health. In May 1883, all of the elevated lines except for the 2nd Avenue line reported operating percentages below 70 percent of gross revenue (table 4.2). These figures illustrate just how profitable a rapid transit business could be in crowded Manhattan with its insatiable demand for efficient transportation.

By definition the elevated railways separated one mode of transportation— rapid movement of a large volume of passengers over longer travel distances—

Table 4.2. Manhattan Elevated Railway Company monthly statement for May 1883

	Railway line				
	2nd Ave.	3rd Ave.	6th Ave.	9th Ave.	All
No. of passengers	48,840	294,176	196,190	52,041	591,247
Total revenue	$49,011	$294,953	$197,490	$52,300	$593,754
Total expenses	$40,650	$128,784	$100,185	$35,285	$304,904
Net earnings	$8,361	$166,169	$97,305	$17,015	$288,850
Operating ratio, %[a]	82.9	43.7	50.7	67.5	51.4

SOURCE: "Manhattan Elevated Financial Statements, 1880–1883," Cyrus Field Papers, Manuscripts Division, New York Public Library, box 4.

[a] Percentage of expenses to revenue.

The 2nd and 3rd
Avenue elevated
railroads at
Chatham Square
in lower Manhat-
tan, circa 1890.

from the crowded city streets. Other traffic, such as horse-drawn freight and slower-moving hackney cabs, traveling to local destinations remained on the streets below. Separating different modes of traffic offered dramatic improvement in travel for passengers riding above the crowded streets. Of course, the elevated trains came with costs. A network of iron and tracks filled the air above the streets, crowding out the light. Steam engines thundering up and down the major north-south avenues created noise, dirt, soot, and smoke. Living next to the Els challenged the nerves of residents, and the tenements lining the avenues next to the Els remained among the least desirable places to live in New York.

The elevated railroads continued to operate well into the twentieth century although they became less and less prosperous due to competition from the subways. During the Depression, Mayor LaGuardia led the call to demolish the elevated railroads, starting with the 6th Avenue El in 1938. The last elevated line running in Manhattan, the 3rd Avenue El between Chatham Square and 149th Street, ended operations in May 1955. Demolition soon followed.

Underground Rapid Transit

Even with the success of the first elevated railroads, the idea of building a subway gained momentum, especially after the opening of the world's first subway in London in 1863. Interest in building subways in New York increased in the 1890s as electricity became a practical means to power rapid transit. The ever-inventive

GRAND CENTRAL'S ENGINEER

electrical engineer Frank Sprague tried, unsuccessfully, to interest the owners of the city's elevated railways in converting from steam to electric power using his patented "unit-operating" system to control multiple electric-powered cars coupled together. Private capital were reluctant to build a subway line, given the huge costs. Building an elevated railroad required $100,000 to $150,000 a mile for construction; the first estimates projected subway construction costs at over a million dollars a mile.

All plans to have the city invest in subway construction ran up against the New York State Constitution's limitation on the amount of debt the city could assume. By law the city could not borrow an amount more than 10 percent of the assessed value of all property. Since New York relied on the property tax as its main source of revenue, the 10 percent limit provided a safeguard against the temptation of elected officials and the political machine to overspend. Even if the city could raise the money to construct a subway, operating it raised additional issues. Could a publicly operated subway in New York not fall under the control of the politicians and be turned into a source of political patronage and corruption? Political corruption on a monumental scale remained a part of the political process in New York, and fears of Tammany Hall stalked any discussion of city spending.

In an attempt to insulate public transportation from the political process, New York State established the Rapid Transit Commission in 1891 to regulate the city's street railways and elevated railroads. The state also delegated to the commission control over all efforts to construct subways. Many citizens believed the five commissioners, appointed by the governor and mayor, would not be subject to political pressures and would therefore bring a rational planning model to address the rapid transit needs of New York. However, a long debate ensued over whether the city of New York should build and operate rapid transit or turn to private companies, as was the model for the nation's railroads.

As early as 1884, Mayor Abram Hewitt had advocated municipal ownership as the only practical means to get a subway built. Speaking seventeen years later, in 1901, Hewitt observed: "It was evident to me that underground rapid transit could not be secured by the investment of private capital, but in some way or other its construction was dependent upon the use of the credit of the city of New York. It was also apparent to me that if such credit were used, the property must belong to the city."[11] Hewitt argued for public financing and ownership but also insisted upon construction and operation of the subway by private contractors. To build a consensus supporting public ownership, Mayor Hewitt and the New York City Chamber of Commerce worked with the legislature in Albany to pass a bill putting the question of public ownership on the ballot for the November 1894 election. On November 7, 1894, the voters overwhelmingly supported municipal ownership of the subways—132,647 for (75.5%) and only 42,916 votes against.[12]

With the voters' approval, the Rapid Transit Commission was confident that private companies would now come forward with proposals to construct the subway with public money and then operate the system as a private enterprise, in a public-private partnership. After protracted negotiations, in January 1900

two companies submitted bids, and the commission selected the firm of John B. McDonald as the prime contractor. The contract (200-plus pages of fine print) required McDonald to immediately deposit $1 million in cash and furnish a surety bond of $5 million to guarantee construction. McDonald failed to raise the needed money, and August Belmont, a New York financier, stepped forward and agreed to form a new company, raise the necessary security, and guarantee the construction of the subway. Belmont, John McDonald, and a group of prominent New Yorkers formed the Interborough Rapid Transit Company (IRT). After receiving approval from the commission, the city, and the state of New York, they signed a final contract on February 21, 1900; and work began.

Lithograph of the crowded City Hall subway station, New York City's first subway, 1908.

The IRT planned the first subway line to start at City Hall, run up to Union Square at 14th Street, and then under Park Avenue north to Grand Central at 42nd Street. At 42nd Street, the line turned west and ran under 42nd Street to Times Square and then north under Broadway to the upper west side. Construction began in March, preceded by appropriate ceremonies at City Hall. Four years later, the first subway train officially commenced service from City Hall Station on October 27, 1904. Completion of the subway line on the west side to the tip of the island and then into the Bronx came in 1908.

Traffic on the subway immediately exceeded all projections. In its first year of operation, the only partially completed IRT carried 106 million passengers. To give a sense of the magnitude of this ridership, in 1900 the population of the city of New York, including the five boroughs with the city's consolidation in 1898, totaled 3,437,202. Manhattan, the only borough served by the IRT in its first year of operation, counted 1,850,093 residents. Thus, total ridership during the first year of operation averaged 57 rides for each resident living in Manhattan. The top five stations in volume that first year were Brooklyn Bridge, Grand Central, 14th Street, Fulton Street, and Times Square. The Times Square station, the *New York Times* reported, was "far ahead of all other local stations. Five million persons, a number exceeding by half a million the estimated total population of New York City, boarded subway trains there during the year. Supposing that just as many got off at Times Square, 10,000,000 persons have passed through the station since Oct. 27, 1904."[13]

Not only did the number of passengers exceed expectations; so did revenue. Since the IRT leased the Manhattan Railway Company, the company's balance sheet included revenue and expenses for both the subway and the elevated railways. For the three consecutive fiscal years beginning in 1906–7, combined subway and elevated revenue increased steadily from $23 million to close to $26 million in 1908–9 (table 4.3). Traffic on the elevated lines remained stable; the real growth in both passengers and revenue came from the subway. The IRT's financial obligation to the city of New York included paying the 3.5 percent interest on the total of $54,802,944 in bonds the city issued to pay for construction of the subway. On the balance sheet, interest payments to the city for the bonds were included in "expenses" and totaled $1,918,103 in 1908. Revenue more than

Table 4.3. Summary of annual reports for the Interborough Rapid Transit Company, 1906–1909: Manhattan Railway elevated railroads and IRT subway

	1906–1907	1907–1908	1908–1909
Gross income	$22,363,802	$24,059,299	$25,775,392
Expenses	9,593,331	10,722,694	10,747,443
Net income	12,770,471	13,335,605	15,027,949
Other Income	815,832	1,220,170	1,384,644
Total Income	13,586,303	14,556,775	16,412,593
Interest on bonds, Manhattan Railway and IRT	4,376,894	5,069,650	5,822,963
Taxes	1,377,965	1,686,466	1,799,807
Balance	7,332,444	7,900,659	8,789,823
Dividend, Manhattan Railway stock	4,116,000	4,200,00	4,200,000
Balance	3,716,444	3,700,00	4,589,823
Dividend, IRT stock	3,150,000	3,150,000	3,160,000
Surplus	$566,444	$550,659	$1,439,823

SOURCE: "Interborough Rapid Transit Co.," *Standard Financial Quarterly*, vol. 1 (New York: Standard Statistics Bureau, 1909), 301.

covered the company's obligation to the city and left a sizable surplus used to pay dividends. Stock dividends in 1904 were 2 percent, but increased to 8.5 percent in 1905 and to 9 percent in 1909, very handsome returns indeed.[14]

Along with the healthy earnings and a tremendous number of riders each day came a great deal of criticism for the IRT and its financier, August Belmont, dubbed the "traction king." Many questioned why the "profits" from operating a city-financed and city-owned subway should accrue to a private company and its wealthy stockholders, who included Belmont, J. P. Morgan, and the grandsons of Commodore Vanderbilt.[15] No one had anticipated just how lucrative the IRT's franchise would prove to be. Criticism of the IRT and Belmont reflected a broader current of populism. Ordinary citizens questioned the accumulation of great wealth, especially wealth generated from the nickel fares of hundreds of thousands of New Yorkers jammed into the subway each day as they traveled to work.

With the stunning success of the IRT, New York politicians, business leaders, the press, and the public clamored for additional subway lines. Developers in the South Bronx and Queens realized what a bonanza the coming of the subway created. When the IRT reached South Bronx, a building boom ensued along the subway route. Construction of thousands of new apartments followed, and New Yorkers, eager to leave the crowded tenements in Manhattan, moved to the Bronx. Between 1900 and 1910, when the first subway line arrived in the South Bronx, the borough's population more than doubled, growing from 200,507 in 1900 to 430,980 in 1910.

As the Rapid Transit Commission moved cautiously forward with plans for an expanded subway system, public pressure increased. On December 14, 1905, the commission held another public hearing; the meeting "was so jammed that those in it could scarcely *move an arm or a leg*."[16] The outer boroughs were well represented. A delegation of 250 Brooklyn residents from Bensonhurst and Bath Beach demanded a subway to their area of Brooklyn. Residents from Ozone Park in Queens wanted changes made to the proposed Jamaica Avenue line to include a loop to provide Ozone Park with direct subway service. Residents in the outer boroughs wanted the subway to be expanded because they faced the same challenges caused by New York's geography that freight shippers and the railroads did. The rivers to the east and north created the same barriers to Manhattan for passengers as the Hudson River to the west. Eventually, subway tunnels under the East and Harlem Rivers dramatically improved the flow of people between the Bronx, Brooklyn, and Queens and Manhattan.

With continued delays in expanding the subway, confidence in the Rapid Transit Commission weakened. When Governor Charles Evans Hughes, a progressive Republican, was elected in 1906, the legislature replaced the Rapid Transit Commission with the more powerful Public Service Commission. The PSC's mandate included more closely regulating all utilities, including rapid transit companies in New York City.

In January 1908, speaking before a real estate group, August Belmont declared that no new subways would be built by private capital unless "those whose capital is invested in transportation were assured a fair return on their investment."[17] Belmont emphasized that the city of New York, with many pressing needs, such as an expanded water system and more schools, did not have the resources to construct and operate a subway system as a public enterprise. If private capital undertook such massive projects, a guaranteed fair return on investment was needed. He then added a veiled threat, stating that "some method would have to be provided by which private capital was properly protected. . . . I wish to say here tonight that if that is not done, you won't see another foot of transportation line built in greater New York." In May, the New York State Legislature passed the Robinson Bill, allowing the city and the Public Service Commission to extend the duration of subway franchises to 20 years, with the possibility of a franchise extending to 35 years, precisely the type of protection of private investment Belmont demanded.

In 1908 the legislature gave the PSC expanded power to negotiate with private companies to expand New York City's underground rapid transit, including Belmont's IRT. Just before the legislature passed the Robinson Bill, the *New York Times* weighed in and urged passage but referred to the numerous plans put forward for new subways as a "perfect Babel of incompetent counsel, by putting forward of impossible schemes of construction."[18]

After fits and starts and numerous revisions, the PSC finally published, in February of 1910, its own plan—the "triborough" system—for expanding the subway by organizing a franchise for a separate, independent subway with lines to Brooklyn, Queens, and the Bronx.[19] The IRT immediately recognized the threat posed by the plan for a competing subway and began efforts to scuttle the proposal. Opposition to the PSC's triborough plan came not just from the IRT and August Belmont. At a meeting of the American Institute of Electrical Engineers, Frank Sprague, one of the members of the Electric Traction Commission for the Grand Central electrification, argued that the planned 2nd Avenue subway down the east side of Manhattan below 42nd Street ignored the west side: "The triborough route south of Forty-second Street parallels the existing subway for a distance of 4 miles within a block on one side or another. It doubles service to Grand Central, but ignores the new Pennsylvania Station."[20] William Wilgus, another discussant at the meeting, echoed Sprague's criticism. In fact, Wilgus and a group of partners had by this time submitted a new proposal for a subway and an elevated railroad to serve lower Manhattan.

The Wilgus Subway and Inter-Terminal Belt Line Plan

By 1908, according to a *New York Times* analysis, "at least 425,866 people go into the business section of Manhattan below Chambers Street every morning and leave at night." With an increasing number of skyscrapers, the *Times* predicted that if all the office workers left their buildings at the same time, "it would require sidewalks in six layers to give them walking space."[21] Over 14,000 commuters arrived by the Staten Island ferries each day, and an estimated 80,000 people crossed to Manhattan over the Brooklyn Bridge. An additional 170,000 people rode the Els and subway to lower Manhattan each day.

All of these commuters created an enormous crush of people during morning and afternoon rush hours. One expert suggested as a solution requiring staggered work hours to spread out the rush hours. All of the experts, including William Wilgus and the PSC, believed that congestion would only increase. The solution to improving circulation lay in expanding the subway system.

On March 9, 1909, on behalf of the "Inter-Terminal Belt Line" syndicate, Wilgus and his Amsterdam Corporation partner, Henry J. Pierce, sent a long letter to the Public Service Commission outlining plans for a network of subways and elevated rail lines around lower Manhattan. The plan proposed connecting all of the rail, shipping, and ferry terminals on the island below 60th Street with a subway and elevated railroad lines along the shoreline. The comprehensive letter of proposal touted the proposed line's convenience for passengers as well as its contribution to solving the freight problem on the west side and asserted that it would help "develop sections of the city, especially on the East Side, which now entirely lack proper transportation facilities."[22]

Because their small-car freight subway proposal was still before the PSC and because the Amsterdam Corporation was in reality a consulting company, Wilgus and Pierce formed a syndicate of wealthy New York investors and engineers to back their new application. Officially, the Inter-Terminal Belt Line syndicate submitted the proposal, with Wilgus and Pierce listed as the principal officers. Their partners included Gustav H. Schwab, Charles Halsey, James White, Charles Cuyler, P. A. S. Franklin, and Anton Hodenphy.[23]

James White, an engineer, had graduated from Cornell University with a Ph.D. in 1885. He headed the J. G. White Engineering Corporation, with offices in New York. He also was serving as director of the Merchant's Association of New York, a business organization very active in promoting improved transportation in the city. White's professional affiliations included the American Institute of Electrical Engineers, the American Society of Civil Engineers, and the American Society of Mechanical Engineers. His involvement in the syndicate added another engineer with a national reputation to the application.

Gustav Schwab came from a prominent German-American family long involved in international shipping; the family firm, Oelrichs & Company, represented the North German Lloyd Steamship Company. Like White, Schwab was active in the Merchant's Association. He also was a director of the United States Trust Company, the Merchant's Trust Company, and the Atlantic Mutual Insurance Company. Schwab's shipping interest and links to the world of New York banking added further credibility to the syndicate. Hodenphy, Halsey, Cuyler, and Franklin all worked on Wall Street; Halsey and Hodenphy headed their own investment and brokerage businesses.

The Inter-Terminal Belt Line syndicate agreement allocated 40 percent of $100,000 in capital of the syndicate to Wilgus and Pierce as the Amsterdam Corporation. All of the other syndicate members shared the remaining 60 percent equally and agreed to cover all expenses involved in the application process, including legal fees.

Before sending the Inter-Terminal proposal to the PSC, Wilgus and Pierce worked behind the scenes to line up political support. Pierce spoke to PSC commissioner Edward M. Bassett by phone on February 13, 1909, and discussed the potential political opposition to their application for a franchise to build the Inter-Terminal Belt Line. Bassett candidly informed Pierce that Governor Hughes would play a key role: "The Governor has felt, up to very recently, that no further rapid transit facilities should be established in New York City except by the city. He is now shaken in his belief and is trying to make up his mind as to the terms upon which franchises should be granted to private capital."[24] Bassett suggested setting up a meeting with the governor, which Pierce arranged for February 25 at the Astor Hotel. Pierce met with the governor accompanied by a number of members of the syndicate to privately lobby Hughes to support their plan.

Bassett told Pierce that the PSC had not received any other application for a subway or elevated railroad along the Hudson or East Rivers. He also indicated that Pierce should not worry about any competing franchise application. In a memo to Wilgus, Pierce summarized Bassett's remarks: "He thought we need not have fears that anyone would get ahead of us because we had more knowledge upon the subject than anyone else, were better equipped and had established a position which could not be dislodged."[25]

Bassett clearly provided Pierce and Wilgus with key insider information from the Public Service Commission. Today, providing such information would violate a code of ethics for a public commission, if not violate the law. In 1909, another ethos prevailed, and neither Wilgus nor Pierce saw any ethical conflict in calling a commissioner to discuss their application for a franchise before they officially submitted the plan to the commission.

Bassett offered one more crucial insight. The Inter-Terminal Belt Line could not be just for passengers. He strongly suggested that on the west side of Manhattan, along the Hudson River, the Belt Line elevated railroad tracks carry both passengers and freight. Adding freight service on the Belt Line would resolve the long-standing "west side problem" and enable the New York Central to remove its tracks from the city streets, eliminating "Death Alley" on 10th Avenue. Pierce reported that Bassett was especially interested in having the Central cooperate and added: "If the Central would make some concessions, the whole matter might be settled to the satisfaction of everyone."[26] As with Wilgus and Pierce's plans for the small-car freight subway along the west side of Manhattan, the participation of the New York Central Railroad remained crucial. If the Central agreed to use the Inter-Terminal Belt Line to move freight, Wilgus, Pierce, and their partners in the Inter-Terminal syndicate would gain the ultimate prize: a franchise from the PSC supported by both the governor and mayor. Once again, looming in the background stood the powerful New York Central tenaciously guarding its west side franchise.

The Inter-Terminal Belt Line

Plans for the belt line included both subways and elevated railroads to encircle the lower part of Manhattan Island from 60th Street on the west side to 59th

Street and the Queensborough Bridge on the east side. The proposal Wilgus submitted divided the project into four major sections:

West Side Section

A four-track elevated railroad forms a connection with the New York Central at 60th Street to the Battery, with intermediate connections at 59th Street with the upper Crosstown Section and at 43rd Street with the Intermediate Crosstown Section. It is the purpose of this section to give direct access to the railroad and steamship terminals on the west side of the Island. . . . It will also be possible to utilize this section for movement of freight trains of the New York Central between the 60th Street and 30th Street yards of that company and to St. John's Park. . . . This arrangement has the merit of affording an immediate solution to the West Side problem so as to permit the removal of the tracks at grade in Eleventh and Tenth avenues and Canal and Hudson streets.

East Side Section

From the terminus of the four-track West Side Section at the point of interchange with the elevated and subway lines and the Municipal Ferry at the Battery, a two-track elevated is proposed along the East River waterfront to a point near 25th Street, where a transition is made to a subway in First Avenue. . . . A junction is proposed with the Intermediate Crosstown Section at 43rd Street.

Upper Crosstown Section

A two-track subway is proposed in 59th Street connecting the Queensborough Bridge and the East Side Section at First Avenue with the West Side Section on the North River waterfront, intermediate stations to be established . . . with the Second and Third Avenue elevated lines, the proposed subway at Lexington Avenue, the possible future subways at Madison and Fifth avenues, the Sixth Avenue elevated line, the future 8th Avenue subway and the existing subway at Columbus Circle.

Intermediate Crosstown Section

For connecting the East and West Side Sections with the Pennsylvania Railroad and Grand Central stations and the Times Square station of the existing subway, a crosstown subway is proposed connecting with the West Side Section at 30th Street . . . under 30th Street to Eighth Avenue . . . northerly under Eighth Avenue to 43rd Street; thence easterly under 43rd Street to a connection with the East Side Section at First Avenue.[27]

Wilgus rested the proposal on a number of underlying principles. The Inter-Terminal Belt Line would charge five cents for a passenger to ride over the entire line. The franchise would have to be for a sufficient period of time so that private investors "will be made as secure, as to return of interest and principal." Once revenue paid all operating expenses, interest, and principal on bonds and depreciation, the syndicate proposed to share profits with the city. At the end of the franchise period, the city could choose to purchase the Inter-Terminal at "fair value" or to extend the franchise for another period of years. A key premise assumed the city would provide the right-of-way over and under the streets free of cost.

One crucial part of the plan involved Manhattan's west side. With the elevated railroad carrying passengers connecting to the ferries and steamship terminals lining the Hudson River and the freight traffic of the New York Central Railroad, the plan solved the long-running west side problem. Wilgus knew that neither the city nor the PSC would approve a franchise for an elevated railroad along the Hudson that did not include the removal of the Central's tracks on the city streets. Adding an elevated railroad to the west side and leaving the Central's tracks on the city streets only exacerbated the problem.

From a broader perspective the west side problem included more than just the New York Central's freight trains on the city streets. All along the waterfront, steamship company piers poured significant traffic onto West Street and adjacent city streets. A passenger arriving by ship or coastal steamer and then transferring to the 9th Avenue El walked three or four blocks east to the most convenient station on the 9th Avenue El. In addition to the steamship lines, thirteen railroad passenger ferry lines from New Jersey docked along the Hudson below 60th Street. The New Jersey–based railroads operated the ferries for the convenience of their long-distance passengers, but large numbers of commuters also rode the ferries across the Hudson each day to work in Manhattan. If ferry passengers wanted to transfer to the Els, they walked a number of blocks east to 9th Avenue. Ferries carried a mammoth number of passengers back and forth each

Table 4.4. Passengers carried by Hudson River railroad ferries, 1908

Rail line	Route (origin to Manhattan)	Slips	No. of passengers	
			1908 total	Daily avg.
Central of New Jersey	Jersey City to Liberty St.	2		
	Jersey City to 23rd St.	1	14,618,405	40,050
Erie	Jersey City to Chambers St.	2		
	Jersey City to 23rd St.	2	19,309,807	52,903
Lackawanna	Hoboken to Barclay St.	2		
	Hoboken to Christopher St.	2		
	Hoboken to 23rd St.	2		
	Hoboken to 23rd St.	2	41,500,000	113,699
Pennsylvania	Jersey City to Cortland St.	2		
	Jersey City to Desbrosses St.	2	34,945,175	95,740
West Shore (NY Central)	Weehawken to Cortland St.	1		
	Weehawken to 42nd St.	2		
	West, NY, to 42nd St.	1	8,823,495	24,174
Total			119,196,882	326,566

SOURCE: *Report of the Public Service Commission*, vol. 2, *Statistics of Transportation Companies for the Year Ended June 30, 1914* (Albany, NY: J. B. Lyon, 1915), 30, table 4.

day and created continuous traffic congestion at the ferry terminals on the Hudson, with numerous ferries arriving and departing during the day and night (table 4.4).

Wilgus argued that even with all of the planned improvements in transportation links between Manhattan, the outlying boroughs, the growing suburban areas, and the long-distance rail links, "no system appears to have been devised for efficiently caring for internal circulation on the Island."[28] An elevated railroad along the west shore of the Hudson, with stations serving the major ferry terminals (Barclay, Cortland, Desbrosses, Christopher, Chambers, 23rd and 42nd Streets) addressed the circulation problem. Passengers arriving by ferry from New Jersey would use the elevated to connect with all of the other mass transit lines in Manhattan.

The Manhattan Shoreline

Over the centuries, the city used landfill to create space for expanded piers and the shorefront streets along the Hudson and East Rivers. With the landfill, the original shoreline of Manhattan that Henry Hudson first saw disappeared, and as commerce increased, the shoreline of Manhattan expanded out into the rivers. The city referred to the streets created along the Hudson shore, including West Street and 11th Avenue, as the "marginal way." Wilgus argued that using the streets immediately inland for an elevated railroad made sense because "the marginal ways were laid out on broad lines in anticipation of a project of this nature."[29] He did not view an elevated railroad along the river as an eyesore. In fact, he predicted an increase of property values with the construction of the Inter-Terminal.

To modern sensibilities, an elevated railway or highway along any shore creates an aesthetic eyesore as well as a physical barrier. An elevated structure acts as a barrier between the city and its waterfront, depriving the people who live in the city from enjoyment of and access to the river or harbor. In San Francisco, the elevated Embarcadero Freeway, built as part of the Interstate Highway System, created a visual and physical barrier cutting San Franciscans off from the waterfront of the Bay. Boston's Central Artery separated the North End and the wharfs on Boston Harbor from the rest of the city. At enormous cost, the Big Dig placed the highway underground and removed the elevated highway that scarred the Boston downtown for decades. In New York, the construction of the elevated West Side Highway produced a similar effect. After long and loud public condemnation, elevated highways came down; and in San Francisco, Boston, and New York the public regained access to the shoreline. By contrast Wilgus's 1909 proposal did not seem out of place, given that the New York City public transit system already included four major elevated railroads running up and down Manhattan Island which millions depended upon for transportation.

On the East River side of lower Manhattan, the proposed Inter-Terminal combined a subway and an elevated railway from the Battery to 25th Street. The piers along the East River, lining South Street, served both international and

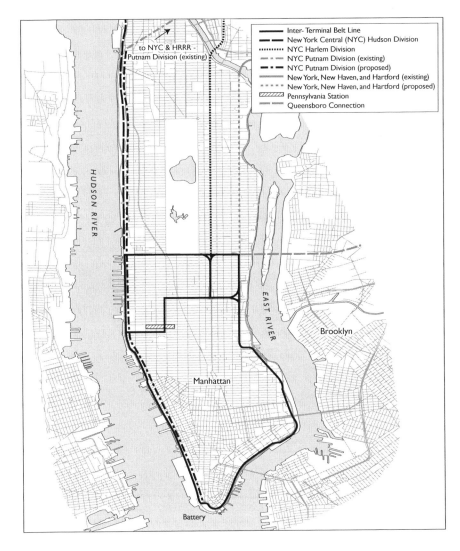

Legend:
- Inter- Terminal Belt Line
- New York Central (NYC) Hudson Division
- NYC Harlem Division
- NYC Putnam Division (existing)
- NYC Putnam Division (proposed)
- New York, New Haven, and Hartford (existing)
- New York, New Haven, and Hartford (proposed)
- Pennsylvania Station
- Queensboro Connection

to NYC & HRRR - Putnam Division (existing)

Wilgus's proposed Inter-Terminal Belt Line, 1909: passenger service to the Battery, with connection from the New York Central Putnam Division in the Bronx, down New York Central tracks to 58th Street, and then on the Inter-Terminal Belt Line from 58th Street down to the Battery.

coastal steamship companies. Old pictures of sailing ships berthed on the piers lining South Street show the majestic bowsprits of the windjammers and more modest vessels extending over a cobbled South Street, almost touching the stores and warehouses just across the narrow street. The famous Fulton Fish Market occupied a section of the east side waterfront at the foot of Fulton Street. To the north, the Manhattan terminus of the Brooklyn Bridge included a busy station serving the cable cars which passengers rode back and forth across the bridge.

A number of busy ferry lines provided service to Brooklyn and Long Island City in Queens. The Union Ferry Company carried 24,942,347 passengers between Manhattan and Brooklyn in 1908. The Brooklyn and Manhattan Ferry Company carried an additional 13,756,722 passengers, and the Staten Island Ferry, operating next to Battery Park, added 10,894,323 passengers (table 4.5). Ferry service across the East River to Brooklyn created the same congestion as

Table 4.5. Passengers carried by Lower East River ferries, 1908

| Ferry operator | Ferry | Route (origin to Manhattan) | No. of passengers | |
			1908 total	Daily avg.
Union Ferry Co.	Fulton	Fulton to Fulton St.		
	Catherine St.	Main to Catherine St.		
	Wall St.	Montague to Wall St.		
	Hamilton	Hamilton Ave. to Whitehall St.		
	South	Atlantic Ave. to Whitehall St.	24,942,347	68,335
Brooklyn & Manhattan	Broadway	Broadway to Roosevelt St.		
Ferry Co.	Broadway–23rd	Broadway to 23rd St.		
	Williamsburg	Grand to Grand St.	13,756,772	37,689
Nassau Ferry Co.	Houston St.	Grand to Houston St.	811,289	2,222
City of New York	South Brooklyn	39th to Whitehall St.	1,847,041	5,060
City of New York	Staten Island	St. George, S.I., to Whitehall	10,894,323	29,847
Total			52,251,772	143,156

SOURCE: *Report of the Public Service Commission*, 2:30, table 4.

did the ferries operating across the Hudson. If anything, the congestion on South Street, where many ferries terminated, exceeded traffic on West Street.

Commuter Rail Service

To further improve internal transportation in Manhattan, the Inter-Terminal Belt Line plan included stations at both Grand Central and Pennsylvania Station connecting to the New York Central, New York, New Haven and Hartford, Pennsylvania, and Long Island railroad passenger trains. Suburban commuters also needed an efficient means to travel from Grand Central and Pennsylvania Station to their places of work.

The new Grand Central provided improved commuter access to midtown because both the Hudson and Harlem divisions of the Central and the New Haven operated all of their commuter service to the one terminal at 42nd Street. When Wilgus planned the new terminal and the underground train yard, he had included a suburban concourse on the lower level with the capacity to serve the three suburban lines. With the plans for the Inter-Terminal Belt Line, he proposed a radical change. Since many commuters worked in lower Manhattan, Wilgus proposed to reroute the commuter trains of the Hudson Division from Grand Central to run from Spuyten Duyvil down the west side of the island, connecting with the Inter-Terminal elevated railway at 60th Street. From 60th Street, Wilgus noted in his proposal to the PSC, commuters "will best be served by the existence of new routes along the waterfront . . . that will enable suburban trains . . . to proceed direct to the Battery."[30]

From the perspective of the twenty-first century, this proposal seems prepos-terous. After all, when Commodore Vanderbilt opened his Grand Central Depot at 42nd Street in 1887, he decided to send all Hudson River Railroad passenger trains to the east side of Manhattan, where once they had traveled down the west side all the way to lower Manhattan to a passenger terminal at St. John's Park. When Wilgus presented the Inter-Terminal plan in 1909, the opening of the new Grand Central was four years in the future and the midtown business district around 42nd Street awaited further development. Most commuters to 42nd Street did not work in the immediate area but, rather, traveled to lower Manhat-tan for work in the city's commercial district. Grand Central served as a transit point where passengers transferred from the railroad either to street railways, elevated railroads, or the newly opened IRT subway to complete their journey downtown. Wilgus assumed the New York Central would find the opportunity to offer service directly to lower Manhattan instead of to Grand Central attractive.

Plans for the Inter-Terminal Belt Line also included a subway connection to the new Pennsylvania Station then under construction at 33rd Street and 7th Avenue. From the moment the Pennsylvania Railroad secured state and city ap-proval for its station and announced construction, it had applied as much pres-sure as possible to secure a rapid transit connection to the station. While the 6th and 9th Avenue Els had long been in service, they were each a block away from the new Pennsylvania Station. The nearest IRT subway station was a number of blocks away, across town on Park Avenue.

The railroad achieved a coup in securing the rights to bring its passenger service onto Manhattan Island. The magnificent station at 33rd Street opened a year before its archrival, the New York Central, finished the new Grand Central. But all of the enormous expenditures for Pennsylvania Station would not lead to parity with the New York Central's service to Midtown if Pennsylvania Station lacked a direct connection to the city's rapid transit system. The railroad argued in the press and privately to the city's political leadership that any expansion of the subway absolutely must include a subway line and station to serve Pennsyl-vania Station. The railroad went so far as to threaten to oppose any plan for expanding subway service that did not include a new station at 33rd and 7th. Wilgus understood the need for subway service to Pennsylvania Station and in-cluded a station on the cross-town section of the Inter-Terminal Belt Line to serve the new station.

During private negotiations with the New York Central to use the Inter-Terminal elevated tracks on the Hudson to move freight, Wilgus proposed creating another link for the Central's commuter service to Westchester County. Today, MetroNorth operates both the Harlem and Hudson divisions with commuter service to the western side of Westchester County along the Hudson River and to White Plains and Brewster, to the east. For many years a third railroad line served Westchester County. Originally built as the New York & Boston Railroad and chartered in 1869, the railroad ran north from Highbridge in the Bronx, to Briarcliff Manor, York-town Heights, and Brewster in the center of Westchester County. A convoluted corporate history followed. In 1878, the New York City & Northern Railroad leased the New York & Boston and built an extension south of Highbridge with a bridge across the Harlem River and a station in Manhattan at 155th Street and 9th Avenue.[31] At 155th Street, passengers transferred to the 9th Avenue El.

As did many railroads, the New York City & Northern Railroad failed during the panic of 1893. J. P. Morgan, the major bondholder, reorganized the line and formed the New York & Putnam Railroad. Morgan then leased the line to the New York Central in 1894, and the Central operated the line as its Putnam Division.

Far less lucrative than the Central's other two Westchester divisions—the Hudson and the Harlem—the Putnam Division never attracted a large number of commuters. Passengers either transferred at Highbridge to reach Grand Central or ended their commute at the 155th Street Station and then rode the 9th Avenue El, a long way down to lower Manhattan. In 1903, the Putnam Division was operating only five morning commuter trains to 155th Street. One of those trains left from Brewster at 6:55 a.m. Stopping at sixteen suburban stations, the train did not reach 155th Street until 8:55, a long two hours after leaving Brewster.[32] With this level of service, Westchester commuters preferred either the Hudson or Harlem divisions, which provided more convenient and much faster service directly to Grand Central at 42nd Street, 113 blocks to the south.

Wilgus and Pierce included another innovation in the Inter-Terminal Belt Line proposal to gain the involvement of the New York Central. Writing privately to W. C. Brown of the Central in April 1909, while their application for the Inter-Terminal was still before the PSC, they pointed out the disadvantages the Putnam Division faced. Operating with steam engines and terminating at 155th Street, far uptown, the division was a persistent money loser. Wilgus proposed building a railroad tunnel at 158th Street to connect the Putnam with the Central's tracks along the Hudson River. "What we have in mind is a connection by means of a tunnel beneath say 158th Street (. . . three-quarters of a mile), and down New York Central tracks to 60th Street and then on to the Inter-Terminal elevated along the Hudson."[33] To further entice the Central, the Inter-Terminal syndicate proposed to pay the cost of the tunnel and the electrification of the entire Putnam Division all the way to Brewster.

With the tunnel in place, the Putnam trains would transfer at 158th Street, via the new underground tunnel, to the Central tracks on the Hudson at 60th Street, and then travel on the elevated tracks of the Inter-Terminal to the Battery. The Central would then be able to offer commuter service from Brewster, through Westchester County, on to Manhattan Island, and all the way to the tip of the island at the Battery.

Never one for a modest, incremental approach to solving New York's transportation problem, Wilgus proposed a plan to bring commuter trains to the very tip of Manhattan Island. Such a rail connection even offered a decided improvement over service to Grand Central if the commuter's final destination was the booming commercial district at the tip of Manhattan Island below Chambers Street. Wall Street executives with homes in Westchester County would be able to ride a Putnam or a Hudson division train directly to the Battery, a few short blocks from Wall Street, then and now the center of the country's financial system.

Cost Estimates

Building adequate transportation for Manhattan and New York City was always expensive, and the proposed Inter-Terminal Belt Line would be no exception. The initial plan presented by Wilgus and Pierce to the PSC estimated the cost for the Inter-Terminal at a total of $50 million, marginally higher than the city's cost for the construction of the IRT, including the extension to Borough Hall in Brooklyn through the tunnel under the East River at the Battery (table 4.6). But an important distinction of the Inter-Terminal Belt Line plan was that Wilgus and the syndicate proposed to pay for the entire cost of construction in return

Table 4.6. Wilgus's table of estimated costs for the Inter-Terminal Belt Line

West Side Section:		
20 miles of single track on elevated structure at $500,000	$10,000,000	
10 stations at $100,000	1,000,000	11,000,000
Upper Crosstown Section:		
½ mile single track on elevated structure	300,000	
6 miles of single track in subway	4,800,000	
5 stations	700,000	5,800,000
Intermediate Crosstown Section:		
2 miles single track on elevated structure at $500,000	1,000,000	
5 miles single track subway at $800,000	4,000,000	
3 stations	600,000	5,600,000
East Side Section North of 42nd St:		
2 miles single track in subway at $900,000	1,800,000	
1 station	200,000	2,000,000
Total for elevated track and subway system		*$24,400,000*
Equipment		
Cars—330 motor and trailer cars at average price of $10,000	3,000,000	
Track—electrification and signals for 40 miles at $35,000	1,400,000	
Shops and storage yard	1,000,000	5,700,000
Miscellaneous		
Real estate, engineering, interest, legal		9,900,000
Total cash requirements for initial sections		*$40,000,000*
Lower East Side Section South of 42nd Street:		
8 miles single track on elevated structure at $500,000	4,000,000	
1 mile of single track in subway	2,000,000	
150 cars at $10,000	1,500,000	
Electrification and signals	350,000	
Real estate, engineering, legal	2,150,000	10,000,000
Grand total cash requirements for all sections		*$50,000,000*

SOURCE: "Estimated Cost," William J. Wilgus Papers, Manuscripts Division, New York Public Library, box 47, folder K.

Table 4.7. Wilgus's estimates of revenue from the Inter-Terminal Belt Line

100,000,000 passengers @ 5 cents/ride		$5,000,000
Operating—40% of gross passenger		$2,000,000
Operating Income		$3,000,000
Income:	New York Central freight	$500,000
	New York Central passengers	200,000
	Mail/express	275,000
	Total	$3,975,000
Deductions:	Depreciation	$300,000
	Interest on $45 million	2,250,000
	Bonds at 5%	25,000
	Net	$1,400,000
Surplus:	½ to City of New York	$700,000
	¼ to bondholders	350,000
	7% to $100,000 capital	7,000
	Net profit to proprietors	$343,000

SOURCE: "Estimated Cost," William J. Wilgus Papers, box 47, folder K.

for the franchise. The Interborough company had made no similar offer to build the city's first subway line. On the contrary, the city of New York borrowed the construction cost, and the Interborough in turn paid for the equipment and operating costs. The Inter-Terminal syndicate proposed not only to pay for all equipment and operate the line but also agreed to pay the entire cost of construction, something Belmont and the IRT refused to do in their negotiations with the city to expand the subways.

As with all of his projects, Wilgus prepared revenue estimates justifying the scope of the project, especially since the syndicate planned to finance the construction and not rely on the city of New York to sell bonds to cover construction costs. The model assumed 100 million passengers paying five cents a ride. Revenue, totaling $5 million a year, would cover all costs and debt payments and would provide a surplus to be shared between the city and the syndicate (table 4.7). Wilgus estimated a net annual profit of $343,000 to the "proprietors"—the Inter-Terminal Rail Line syndicate. Wilgus and Pierce stood to receive $137,200 (40%), with the remaining 60 percent shared by the rest of the syndicate. If the syndicate was successful in securing a franchise and building the belt line, Wilgus stood to become a rich man.

The projected volume of 100 million passengers a year assumed that the Inter-Terminal would draw many passengers using the Hudson and East River ferries and the East River bridges. In 1908, the Hudson River ferries carried 119,196,882 passengers (table 4.4) and the East River ferries carried 52,251,772 (table 4.5)—a total of over 171 million potential customers. Add to that the number of railroad passengers arriving and departing at Grand Central Terminal and Pennsylvania Station. In addition, the Inter-Terminal Belt Line created an alternative to traveling on the elevated railroads, which carried over 282 million riders in 1908.[34] If even a small fraction of the people traveling on the Els used the Inter-Terminal as an alternative, the estimate of 100 million passengers did not seem excessive. After all, when the IRT opened in 1904, it carried 72 million riders in its first year, and just two years later the total reached 238 million. Demand for rapid transit on Manhattan Island seemed to be insatiable and the projections for ridership on the Inter-Terminal reasonable.

To further refine estimates of passenger traffic on the Inter-Terminal, Wilgus calculated "passengers per mile" per year on each of the four elevated railroads in Manhattan for selected sections (table 4.8). On a yellow pad he broke down passenger volume for each elevated line into three or four sections. Passenger traffic on the Els varied by area of the city; he wanted to develop a model for the area below 59th Street, where the Inter-Terminal planned to compete directly with the Els for passengers.

To put these passenger totals in perspective, the 1900 U.S. Census officially counted the population of the entire country at 76,212,168. On the 3rd Avenue El, the number of passengers carried each year between 1907 and 1909 averaged

Table 4.8. Wilgus's estimates of number of passengers per mile on selected sections of elevated railroads in Manhattan, 1907–1909

Railway line and section		Length (miles)	Avg. no. of passengers	Ticket sales per mile per year
2nd	Chatham to 1st Street	1.04	11,318,058	$10,900,000
	8th to 57th	3.04	11,034,908	3,600,000
	65th to 127th	3.23	19,210,544	6,000,000
		7.31	41,563,510	5,690,000
3rd	Whitehall to Houston	3.85	31,011,428	10,800,000
	9th to 59th	2.85	28,277,864	10,000,000
	67th to 125th	3.12	30,873,295	9,900,000
	129th to Bronx Park	5.50	25,044,061	6,370,000
		14.32	125,206,648	8,750,000
6th	Rector to Bleecker	1.90	15,815,892	8,320,000
	8th to 59th	3.80	30,938,614	8,150,000
	66th to 155th	4.75	38,508,210	8,100,000
		10.45	85,262,716	8,100,000
9th	South Ferry to Desbrosses	1.71	13,785,405	8,505,000
	Houston to 14th	0.95	5,837,833	6,150,000
	23rd to 59th	2.08	11,148,455	5,360,000
		4.74	30,771,693	6,500,000

SOURCE: "PSC data 4 elevated railroads ticket sales," William J. Wilgus Papers, box 47, folder H.

over 125 million, the equivalent of the entire population of the country riding the 3rd Avenue Elevated 1.6 times. Clifton Hood, in his history of the New York subways, drew a similar analogy to illustrate the sheer magnitude of the number of people riding mass transit in New York City: "If the subways and Els were used to full capacity for a twenty-four hour period, they could have accommodated 35 million people, one-third of the nation's entire population."[35] The Public Service Commission, in its annual report, calculated ridership on the city's street railways, elevated railroads, and the IRT on a per capita basis. In 1903, before the IRT subway opened, the street railways and Els carried over one billion riders, 265 rides per capita per year. Two years later, with the first subway running, ridership rose to 305 rides per capita and in 1909 to 321.[36]

In his revenue and cost model, Wilgus included almost another million dollars in revenue from the New York Central's use of the elevated railroad along the Hudson for its freight service to lower Manhattan. The estimates also included the Central's diverting some of its suburban passenger trains from Grand Central to its tracks along the Hudson River and then over the Inter-Terminal elevated railroad to the Battery. A crucial premise of the entire complex plan for the belt line assumed the cooperation of the New York Central Railroad.

Public Reaction and Political Realities

Amid the press coverage of the battle over expansion of the IRT and the organizing of an independent subway company to extend lines to the other boroughs, the proposal to build the Inter-Terminal did not attract the same attention. The *Evening Sun* in March 1909 briefly reported on the syndicate's proposal to the PSC and described the Wilgus plan as "a radical and comprehensive plan for the solution of the city's transportation problems." The *Sun* added that the Inter-Terminal constituted but one of "an abundance of plans for big traction propositions."[37] The Inter-Terminal did attract favorable coverage in the engineering press, including a very supportive editorial in the *Railway Age Gazette*.[38] Continued discussion of the ongoing battle over the IRT dominated the debate over rapid transit in New York City and overshadowed Wilgus's proposal.

Wilgus wrote to the editor of the *New York Daily Tribune* to explain how the Inter-Terminal Belt Line plan complemented his earlier submission to the PSC for the small-car freight subway.

> The plan for small-car freight subways is proposed as a means for collecting and distributing freight. The "Belt Line Project" is primarily for the purpose of affording first-class rapid transit passenger and express facilities to the waterfront. . . . The ultimate solution of the freight problem of Manhattan in relation to all water and rail carriers of the Port; and the "Belt Line Project" is intended for passengers and express, incidentally offering a quick solution of the "west side" problem. The application for the right to construct and operate the passenger Belt Line is not intended to replace the freight subway application.[39]

He offered the same argument to the editor of the *Railway Age Gazette*: "The Inter-Terminal Belt Line Project . . . is supplement to and is not intended to, in any way, replace the application that has been made for treating the freight problem on Manhattan Island by means of the small-car subways."[40] If the Public Service Commission approved the franchise applications for both the Inter-Terminal Belt Line and the small-car freight subway, Wilgus believed the two systems could dramatically improve both passenger and freight service in Manhattan and in the port of New York.

Problems with the New York Central—Again

Wilgus and Pierce knew that the Public Service Commission would not approve the Inter-Terminal franchise with an elevated railroad along the Hudson unless they gained the cooperation of the New York Central. If the Central agreed to utilize the elevated tracks to transport its freight to lower Manhattan and to the piers on the Hudson, their application stood a good chance of approval. As with the small-car freight subway, all forward progress rested on the cooperation of the Central. As early as March, just after the Inter-Terminal franchise proposal became public, W. C. Brown, the new president of the Central, responded to Pierce's request for support for the elevated railroad along the Hudson in no uncertain terms: "No proposition which does not provide for adequate, continued use by the Company for freight purposes could meet with our approval."[41] Brown added that the Central remained interested in the passenger service on the elevated railroad but did not see how both passenger and freight traffic could be accommodated simultaneously. He also admitted that the railroad remained under enormous pressure to solve the west side problem and get its freight tracks off the city's streets.

Since everything hinged on the participation and support of the New York Central, Wilgus and Pierce drafted a detailed response to the concerns Brown raised. On the elevated section from 60th to 30th Streets between the railroad's two major rail yards, the Inter-Terminal proposed giving the Central complete use of two tracks between the hours of 9:00 p.m. and 6:00 a.m. With no anticipated problems handling freight as far south as 30th Street, the real problem involved the Central's freight service from 30th Street to St. John's Park: six million tons of freight in 100,000 cars passed both ways each year. Wilgus estimated that the Central required "at the outside one train per hour in each direction from 7 to 10 hours each day." According to Wilgus's calculations, twice this amount of freight could be handled on the Inter-Terminal elevated each day without interfering with passenger service. However, he recognized that the Central needed to give up its perpetual right to use the city streets. To try to persuade Brown to at least consider using the Inter-Terminal, Wilgus pointed to the advantages of the Central's offering direct suburban service to lower Manhattan: "Enormous advantages will occur to the New York Central through it being able to transport its suburban passengers from both the Hudson and Harlem divisions without change of cars to and from as far south as possible"—that is, passengers would not have to change to the Els or the IRT at Grand Central in order to reach lower Manhattan.[42]

Brown's response was again blunt: "Our rights are perpetual, we have a large amount of money invested throughout the territory and could not consider any proposition that involved the surrender of our present rights for anything less permanent or dependable." The New York Central believed the railroad held the perpetual rights to use the city streets for freight service to lower Manhattan. Brown admitted that the Inter-Terminal, as proposed, could handle the Central's freight and passenger service. Nevertheless, the railroad needed absolute assurance "that the freight facilities be regarded as of first or paramount importance, that these facilities must be adequate beyond any question of doubt, and the right to use them must be perpetual."[43]

No plan, even one as sophisticated as that devised by the Central's former fifth vice president, could persuade the railroad to surrender what it viewed as its absolute right to use the city streets for its freight trains. Over the decades, the Central had invested a great deal of money and earned substantial revenue from hauling freight along the Hudson even as the railroad's freight trains on the city streets perpetuated congestion on the west side. The railroad adamantly refused to consider any plan that threatened its freight service monopoly, regardless of the opposition of the politicians, the press, and the public.

The Ongoing Political Process

As Wilgus and Pierce continued to negotiate privately with the New York Central, important political developments unfolded in Albany. For the Inter-Terminal proposal to succeed, changes to the current rapid transit law were required. The most important change: to allow private companies, rather than the city of New York, to pay for the construction of new subway or elevated lines. After construction the private companies would transfer ownership to the city in exchange for long-term operation franchises, up to 100 years.

The PSC supported the idea of private construction and operation of any new subway lines in New York and actively pushed for enabling legislation in Albany. Pierce, writing to the Inter-Terminal syndicate's lawyer, R. C. Shepard, in early April of 1909, pointed out the importance of the pending legislation: "The Public Service Commission will probably introduce a rapid transit bill in Albany within a few days. Under the terms, if passed, they will be able to negotiate with us; but should it not become law, no further progress can be made in the matter before next year."[44] With the support of the reform-minded Governor Hughes, the legislation passed on April 24, 1909. Newspapers reported the new transit law in favorable terms. The *New York Herald* pointed out the key advantage of the law: to "stimulate subway construction" with "the corporation and the city to share dividends while roads remain under private control."[45]

With the legislation signed by Hughes on May 25, 1909, Wilgus and Pierce believed the PSC stood ready to approve their application for a franchise as soon as they reached an agreement with the New York Central. Negotiations continued through the summer. Pierce wrote to Ira Place, the Central's attorney, in August trying to persuade the Central to agree to use the Inter-Terminal. He proposed three key parts to an agreement between the Central and the Inter-Terminal syndicate. First, the Central would subscribe to 51 percent of the capital stock of the line, giving the Central effective control. Second, the Inter-Terminal would contract with the Central to handle the railroad's freight and passenger service south of 60th Street for 100 years. Finally, the Inter-Terminal pledged to negotiate a deal with the city: "In consideration of the New York Central Railroad abandoning its tracks at grade through the streets south of 60th Street, the city (by aid of an enabling act to be passed by the next Legislature), as owner of the elevated structure, to agree that the traffic contract between the New York Central and the Inter-Terminal shall be carried out in perpetuity."[46] In other words, Wilgus and Pierce pledged to work the political process to transfer the Central's perpetual right to use the city's streets to the Inter-Terminal's elevated railroad.

A month later, neither Pierce nor Wilgus had heard back from Ira Place, and they had no idea whether Place had discussed their latest proposal with W. C. Brown. Pierce again wrote to the Inter-Terminal syndicate's attorney, R. C. Shepard, to suggest that he follow up with Place and arrange a private meeting with President Brown. By this time matters had come to a head because Pierce had decided to accept a new job in Washington State and sever his connection with the Amsterdam Corporation. Pierce suggested to Shepard that he and Wilgus travel to the Adirondacks or some nearby place to meet with Brown, who was on vacation in upstate New York.[47] With time running out and without an agreement with the Central, Wilgus and Pierce knew the PSC would not move forward with their franchise proposal. All hinged on the cooperation of the New York Central.

To compound matters, the bitter, never-ending battle between the New York Central and the city of New York over the railroad's tracks on the west side streets escalated during the spring and summer of 1909. While Wilgus and Pierce negotiated in private with the Central, political pressure on city officials to do something to solve the west side problem reached a fever pitch. At the very end of the legislative session in Albany, Senator Thomas Grady, a member of Tammany Hall who represented Manhattan's fourteenth district, introduced legislation to affirm the Central's perpetual right to use the city's streets for its freight tracks along the Hudson. All hell broke loose, and New York mayor McClellan vetoed the bill after it passed in Albany on May 17, 1909.[48] A few days later, the Board of Estimate, reacting to the public outcry over the Grady Bill, ordered the New York Central to remove its rails from the city streets in thirty days: "If not, they'll be torn up."[49] The Board of Estimate did not expect the railroad to comply; it expected to take the matter to court and legally determine, once and for all, if the railroad maintained a perpetual right to use the city's streets on the west side for freight service.

While the city waged a battle with the Central in the courts, the Inter-Terminal proposal before the PSC stalled. Pierce wrote to the syndicate members in September and laid out the dilemma in stark terms: "This matter is now before the Public Service Commission . . . but as granting a franchise is largely dependent upon our being able to effect a traffic agreement with the New York Central and as the New York Central is in litigation with the New York City authorities as to their rights to operate cars through the streets south of 60th Street, our matters will have to remain in abeyance until the courts have decided the question. . . . While it is problematical as to whether we will obtain a franchise, yet it is our intention to keep matters alive."[50] In a separate letter to the syndicate members Wilgus reported on an informal conversation with the PSC. The commission informed him that no further action would be taken until the syndicate signed an agreement with the Central. Wilgus reluctantly agreed with Pierce: "Under these circumstances there is nothing for us to do but let the matter rest."[51]

For the second time in just two years, a proposal by Wilgus to solve parts of New York City's ongoing transportation problems had been thwarted. He had advocated the small-car freight system as a means of transporting millions of tons of freight across the Hudson. The Inter-Terminal Belt Line was intended to dramatically improve the internal movement of passengers in Manhattan along the shore and throughout the rest of the island. Both plans came to naught because of the absolute intransigence of the New York Central Railroad. The railroad, in its own narrow self-interest, refused to give up the use of city streets along the Hudson. No matter the brilliance of Wilgus's plans, the Central's opposition doomed them. Time, however, provided Wilgus with a measure of vindication.

New Solutions and Challenges
The Expansion of the Subway: The Dual Line

Plans, negotiations, and political battles over the expansion of the subways continued for years. The PSC's triborough plan of 1910 proposed a new subway line more than double the length of the IRT. After much criticism, the PSC finally solicited bids for the construction of the Triborough in the summer of 1910. When the commission opened the bids in October, it discovered that not one private company offered to build the subway with their own capital. Numerous bidders offered to equip and operate the Triborough if the city of New York financed the construction. On the other side, conservative business leaders and the public opposed any more municipal financing.

Wilgus did not stand idly by. In December 1910, A. Barton Hepburn, president of the Chamber of Commerce, and Henry Towne, president of the Merchants' Association, formed the Citizens Committee to advise Mayor Gaynor on the expansion of the subway. The committee included former reform mayor Seth Low and prominent bankers: Francis Hine, president of First National Bank and William Porter, president of Chemical National Bank. The only engineer invited to serve on the committee was William J. Wilgus.[52] Given his frustration over the failure of his Inter-Terminal Belt Line plan to move forward in the face of New York Central's intransigence, Wilgus could hardly have been objective. The committee issued a report within the month that recommended extending the IRT rather than building a competing subway system not integrated with the existing one. While the committee was highly critical of the management of the IRT, its members recognized that having one integrated system with a single fare better served the citizens of the city.[53]

Debate over expanding the subway dragged on into 1911. A new committee chaired by Manhattan Borough president George McAneny recommended a dramatic expansion of the subway system with the participation of the city and both the IRT and the Brooklyn Rapid Transit Company (BRT). Known as the "dual system," McAneny's plan envisioned expanding the subway to include areas of the Bronx not served by the IRT as well as outlying areas in Brooklyn and Queens. With the city of New York providing construction financing and the IRT and the BRT paying for equipment, both companies agreed to share revenue with the city.

Almost all of the city's major newspapers supported the dual system plan. The *New York Daily Tribune* stated, "The city will be in a strong position if it adopts the report."[54] The *Globe and Commercial Advertiser* recognized the truly innovative nature of the plan: "The McAneny committee is to be congratulated on the largess of the plan it has developed. . . . The prospect now is not merely for rapid transit expansion, but for the settlement of the rapid transit problem for a generation. . . . New York will be on a level of Paris and London."[55] The *Times* praised the plan and urged action: "Never has any city work been more exhaustively discussed, and never up to this moment has a better plan been proposed by anyone. Unless talk is to be interminable, and the dirt never fly, here and now is as good a time to turn talk into action as we are likely to ever see."[56]

In March 1913, almost a decade after the first subway opened, the city finally began construction to expand the subway system dramatically, adding 619 miles. As Hood summarizes, "By 1920, New York would surpass London and rank as the largest rapid transit system in the world."[57]

While New Yorkers rejoiced at the plans for the expansion of the subway system, the planned subway lines did not include service along the shores of Manhattan Island. On the west side, the subway stations would still require a walk of several blocks from the piers and ferry terminals on the Hudson. Along the East River a similar situation prevailed. Wilgus continued to argue for the Inter-Terminal plan, including constructing rapid transit lines along the shore of Manhattan Island on both the east and west sides. The huge dual system also failed to include service to the west side of Manhattan for ferry passengers from New Jersey.

The West Side Problem: The High Line

The dramatic expansion of the city subways did absolutely nothing to solve the west side problem. The New York Central continued to use city streets to run its freight service to lower Manhattan and the piers on the Hudson River. The long-running battle between the city of New York and the New York Central Railroad

High Line elevated railroad tracks, 1934, looking north from Bank Street with Washington Street in foreground. The tracks run through the Bell Telephone Laboratory building.

over the use of the city's streets for freight service persisted through the next two decades, even as the major expansion of the subway system moved forward. Finally, in July 1929, the state and city of New York and the Central agreed to a major plan to remove the railroad's tracks and freight trains from the city streets by constructing an elevated freight railroad from the Central's 30th Street yard south to a new St. John's Park freight terminal. Soon to be called the High Line, the elevated tracks ran across private property through the middle of blocks south of 30th Street. Twenty years earlier, Wilgus and the Inter-Terminal Belt Line syndicate had proposed almost exactly the same solution for the freight problem on the west side.

To be fair, the West Side Improvement project, as construction of the High Line was called, included eliminating 105 street-level railroad crossings from

St. John's Park to Spuyten Duyvil and adding 32 acres to Riverside Park by building over the railroad tracks bordering on the Hudson River. With an estimated cost of $150 million, the agreement called for the city and state of New York and the Central to share in the cost of all of the improvements, including the High Line.

The High Line's electrified, elevated tracks ran right through buildings along the route south of 30th Street. Freight cars delivered directly to platforms in the buildings. New buildings constructed above the High Line made use of air rights over the tracks, just as the apartment and office buildings did along Park Avenue north of Grand Central. At 30th Street a spur from the High Line extended to a new post office building at 29th Street. Further south at Spring Street, the new St. John's Park terminal building opened in June 1934. The new building included indoor space for loading and unloading 150 delivery trucks simultaneously, decreasing congestion on the surrounding streets.[58]

The High Line removed the New York Central's freight trains from the city streets, but the railroad continued its monopoly of direct freight service to the island. The expensive investment did not in any way facilitate the delivery of freight from the railroads on the New Jersey side of the Hudson. Thus, the incredible congestion at the freight terminals and piers along the Hudson remained.

Shippers to New York continued to pay a hidden tax on the movement of every ton of freight back and forth across the Hudson. Over the decades, the New York Central's tenacious defense of its perpetual right to use the city streets paid off. On through the 1930s and 1940s, the railroad maintained its monopoly on direct freight service to Manhattan, with its freight trains now running under Riverside Park and then further south on the High Line to lower Manhattan. While the Central benefitted, Manhattan and the port of New York remained without an integrated rail system for freight delivery. By comparison, both Wilgus's small-car freight subway and the Inter-Terminal Belt Line had proposed a more comprehensive system for moving freight. The New York Central's refusal to cooperate worked to its immediate advantage, but the public interest suffered.

"The Highway and Truck Cometh"

In spite of the Central's apparent victory in staving off any competition, truly momentous changes to the city's transportation system began. Because the High Line ran inland from the Hudson River, the marginal way and West Street directly adjacent to the river were open space ripe for highway construction. The shoreline of Manhattan provided the only space on the crowded island for limited-access, high-speed highways. With the automobile and truck age well under way in the New York region, motor vehicles increasingly filled the streets of New York and every other city in the country. Ominously for the railroads, trucks captured more and more of the freight delivery business—first, short haul within the cities and metropolitan regions, and then, with the building of a colossal interstate highway system, across the United States.

Opening of West
Side Highway,
1930, looking
north from
Spring Street,
with Hudson
River piers to
the left.

In 1929, the very year the Central railroad agreed to the West Side Improvement Plan and the High Line, the city of New York decided to build an elevated highway down the west side of Manhattan. Two years earlier in 1927, the opening of the Holland Tunnel had provided the first vehicular tunnel under the Hudson River. The West Side Highway (first named the Miller Elevated Highway, after Henry Miller, the Manhattan Borough president who pushed for construction) used the same construction principle as the elevated railroads. Above 10th Avenue and West Street, steel trestles carried the highway above the city streets with numerous on and off ramps. The opening of the highway allowed vehicles to drive from upper Manhattan to the tip of the island without any stoplights and at much greater speed than on any of the north-south avenues.

While the long, tortuous debate over public versus private ownership of the subways continued, only one model for the new highway systems emerged: public financing and public ownership. New York City, New York State, and cities and states across the country embarked on a frenzy of highway construction. At great public expense, a vast highway system would eventually be constructed across the entire country. Across the country the railroads faced competition from a publicly subsidized mode of transportation, a competition they inevitably lost. Over the next half century all of the mighty railroad companies, the largest companies in the country in 1900, faced utter ruin; bankruptcy followed.

In New York, Robert Moses began his spectacular rise to power, soon to gain control over most of the highway and bridge construction that so changed the metropolitan region. As car and truck traffic increased with the opening of each new highway, bridge, and tunnel, a seemingly inexorable logic demanded more highway construction. The private railroad companies, whether the once powerful New York Central or the much more modest street car companies, could not survive. Local governments still expected the railroads to pay property taxes for each acre of land their tracks and rail yards occupied, while next door a new publicly financed highway provided the private car and truck with a comparatively "free" infrastructure. It is a wonder the mass transit companies managed to limp along for as long as they did while trapped in a downward death spiral. As they lost passengers to the automobile and freight revenue to the trucking companies, service deteriorated, costs increased, and one after another, new bridges and highways opened, accompanied by great public fanfare and acclaim as beaming politicians joined Robert Moses at yet another ribbon-cutting ceremony.

Soon, vehicles using the Holland Tunnel and the West Side Highway delivered more people and freight in the city. As the highway age gained momentum, New Yorkers embraced highway and bridge construction as the future. In the fateful year of 1929 as construction of the High Line began, a long article in the *New York Times* breathlessly described plans to construct more highways like the West Side. No longer would transportation planning focus just on the railroads and subways. Now the traffic problem—the "city's illness" according to the *Times*—no longer involved the omnibus, the street railway, the elevated railroad, the subway, or freight tracks, but rather, cars and trucks jamming the narrow city streets, originally laid out in 1811. The solution proposed: "Vast Projects to Relieve City's Traffic Jam . . . Tri-borough Bridge and Narrows Tunnel . . . Super-Highway System."[59] With the construction of the West Side Highway and the FDR Drive, highways rather than mass transit now encircled Manhattan Island. Private cars and trucks crowded the shoreline rather than the combined subway and elevated railroad proposed by William Wilgus.

While the New York Central enjoyed initial success with the High Line, the railroad fought a long, losing battle for freight service in Manhattan to the truck and the highway. Traffic declined over the High Line, and the freight stations along the line in lower Manhattan were torn down in the 1960s. All freight traffic ended in 1980. Parts of the High Line remained abandoned and deteriorated until park and open-space advocates formed the Friends of the High Line in 1999. These advocates argued for the line's remaining section to be renovated as open space. In July 2003, the city committed over $15 million to a High Line park and open space which opened in 2009.[60]

Wilgus remained convinced of the value of improving rapid transit connections to the Hudson River waterfront, which the Inter-Terminal plan proposed to accomplish. But once again his innovative ideas for improving freight and passenger circulation in Manhattan had proved to be far ahead of their time. Again, he encountered absolute intransigence on the part of the railroads, especially his former employer, the New York Central. Because of a lack of buy-in from the Central and the other railroads, the PSC halted his application for the Inter-Terminal Belt Line. With no relief in sight to ease congestion along the Hudson River piers, shippers turned to trucking companies, using the Holland Tunnel, and eventually the George Washington Bridge and Lincoln Tunnel, to move freight between Manhattan and the New Jersey rail yards, and the truck and the automobile age gained momentum.

5

WORLD WAR AND IDEAS FOR A NEW YORK– NEW JERSEY "PORT AUTHORITY"

DESPITE HIS DISAPPOINTING failure to obtain a franchise for either the small-car freight subway or the Inter-Terminal Belt Line, Wilgus found success consulting for the railroads in Philadelphia, St. Louis, Toledo, Georgia, Chicago, and elsewhere throughout the country. He found that the work both challenged his engineering skills and rewarded him financially, describing it as "my new born freedom" after his bitter departure from the New York Central Railroad.

Across the United States, the several railroads that served most major cities created a jumble of separate freight yards and terminals. Each railroad operated its own freight service, often without any direct connection to the other rail systems. Any freight to be exchanged required unloading at one yard, hauling by dray or freight wagon across town to the rail yard of another railroad, and then loading onto a second freight car—an inefficient and costly process. The movement of freight on city streets between rail yards added to traffic congestion on already crowded city streets.

A belt line railroad allowed freight cars from one railroad to be switched to the tracks of another railroad, eliminating the need to load and unload. In a number of cities, belt line railroads, jointly owned by the major railroads serving a city, shared costs and profits. Rapid interchange of freight increased efficiency and reduced costs. Wilgus strongly advocated constructing belt line railroads in major metropolitan regions. The city of Philadelphia retained him to aid in preparing contracts for a belt line to connect all the railroads to the south of the city.

While Wilgus's applications for a small-car freight subway for delivering freight from New Jersey to lower Manhattan and the Inter-Terminal Belt Line for passengers were still before the Public Service Commission, and while his private consulting practice continued to flourish, he returned to the problems of the port of New York. He also developed a number of plans for a belt line railroad in the New York metropolitan region.

Wilgus's Plan for a Two-State Port Authority

Wilgus realized from his first days in New York working with the New York Central that the transportation challenges of the port of New York involved more than just traffic and congestion on Manhattan island. The city of New York alone could not solve the problems of moving people and freight back and forth across the harbor or of delivering the necessities of daily life for an ever-growing population. Any solutions required the cooperation of both the two states of New Jersey and New York, the numerous municipalities on both sides of the Hudson River, and the borough of Brooklyn.

The fragmentation of political power in this region stymied any comprehensive plan to improve transportation in the port: a bold regional approach was needed. Wilgus had been thinking about the idea of a regional New York–New Jersey port authority for some time and first presented his ideas in a letter of December 29, 1909, to the secretary of the New York Barge Canal (Erie Canal) Commission, Alfred R. Smith: "I am formulating the idea that I mentioned to you of a Metropolitan District."[1] A long memo for his own files of the same date outlined his ideas for a regional New York–New Jersey port authority. He began by pointing to the recent establishment of a metropolitan district in the Boston area to coordinate the water, sewerage, and parks for the entire region. He then drew a parallel: "Perhaps through some similar agency a solution may be found for the many problems that vex the vast population that is immediately tributary to the Port of New York."[2]

Wilgus was a strong advocate of rational planning. All of his training and experience reinforced a certainty that any problem, no matter how complex, could be solved if the necessary data could be assembled and analyzed. Out of this analytical process, a logical plan would emerge. With a plan in place, Wilgus always believed the underlying logic would convince all of the myriad interests—railroads, shippers, local governments, and politicians at all levels—to cooperate. The greater good for the whole would triumph over narrow self-interest.

He proposed an "Interstate Metropolitan District" to include one hundred cities, towns, and villages in the bi-state region, with an aggregate population of six and a half million people surrounding the busiest port in the United States and the world. To solve the transportation problem Wilgus prepared a detailed list of needed improvements: an outer freight belt line entirely encircling the port connecting with existing and future intersecting railroads, an interior line for freight distribution with a freight subway in Manhattan modeled on his "small-car" subway proposal from 1908, and waterfront terminals at suitable locations with railroad connections and ample piers and warehouses.[3] He also insisted on a centralized system of operations so that the freight cars of all of the railroads could reach any part of the port, without discrimination, and at the same rate.

Wilgus envisioned creating the "greatest aggregation of human beings in the world under a centralized civic government." One comprehensive government agency would exercise power for the common good of all. He added a proviso that each local government "would preserve its autonomy" but would follow "general conditions and regulations" established by the district, guided by a panel of transportation experts who would "arrive at its conclusions only after an exhaustive study."[4] The Metropolitan District would exercise power over the entire port

Table 5.1. Wilgus's comparison of proposed Interstate
Metropolitan District in New York with London

Area	No. of square miles	Population
New York Metropolitan District	1,928	6,628,200 (1909 est.)
City of London	120	4,536,063 (1901)
Greater London	690	6,580,616 (1901)

SOURCE: Wilgus, "A Suggestion for an Interstate Metropolitan District," 16–17, William J. Wilgus Papers, Manuscripts Division, New York Public Library, box 48.

of New York with the guidance of experts, combining a controlling authority with rational planning.

Accompanying the outline for the Metropolitan District, Wilgus included a number of maps; the first showed the outer belt freight line connecting all of the railroads serving the port. Wilgus's plan included building a major shipping port in Jamaica Bay and a freight terminal and piers on City Island where Long Island Sound meets the East River in the Bronx. A second map compared the size of the New York metropolitan district with the recently created Boston district to illustrate the much greater size and complexity of the port of New York region. A table compared the New York port district to the City of London and the Greater London Metropolitan District (table 5.1).[5] The London data illustrated that no other metropolitan region as complex as New York existed anywhere in the world with the "novelty and magnitude of the problems, and the enormous expense that they will involve."

Wilgus knew no one political jurisdiction alone could solve the problems of the port of New York. To move the proposal through the political process required joint action by both New York and New Jersey. Wilgus, who thought the project might cost as much as the Catskill reservoir project or the Panama Canal, wrote to various state officials and politicians and called for a detailed study of the port's transportation needs. In another letter to Alfred Smith of the Barge Canal Commission, he spelled out in greater detail the underlying premise: "What I have in mind is not a city commission, but an interstate commission . . . that would be a guide for the actual execution and supervision of the work by an 'Interstate Metropolitan District.' . . . I do not think anything will ever come of a city commission."[6]

In 1910 Wilgus wrote to his friend Gustav Schwab, who served as an officer in the New York Chamber of Commerce. Wilgus again reiterated that no single municipality within the port region could work alone and suggested three concrete actions. First, the New York City Chamber of Commerce should join with chambers in Yonkers, Newark, Jersey City, and other important communities within the port and recommend to the legislatures of New York and New Jersey the appointment of an expert interstate transportation commission. Next, the new commission would then carry out a detailed study of port conditions, and finally, it would provide recommendations for forming a permanent metropolitan transportation district.[7]

These recommendations, proposed by Wilgus in January of 1910, provided a blueprint for the steps taken seven years later with the appointment of the New York, New Jersey Port and Harbor Development Commission. In 1918, the Harbor Development Commission carried out an exhaustive study of the port of New York, drew up a series of specific recommendations for change, and recommended to the legislatures the establishment of a permanent port commission. From this recommendation, New York and New Jersey enacted legislation creating the Port of New York Authority in April of 1921.

Wilgus continued to advocate for a bi-state commission. In a 1910 letter to the editor of *Engineering News* he reiterated his firm belief that only unified, collective action could solve the port's inherent transportation challenges: "No private corporation or group of individuals can alone bring this about." He was putting forth his proposal "with a view to starting a movement that will result in some definitive [action] . . . the appointment of an Interstate Terminal Commission."[8] Again, he pointed to the success of the metropolitan district formed in Boston as a model to emulate. The editor of *Engineering News* responded immediately and asked Wilgus to write an article for publication in the journal. In the March 31, 1910, issue of *Engineering News* Wilgus again clearly laid out his argument for a unified port district: "It may be feasible for New York and the neighboring cities on both sides of the Hudson to have a 'Metropolitan Freight Transportation District,' embracing all parts of the Port. . . . Without joint action, any broadly effective improvement in terminal conditions is hopeless."[9]

While Wilgus emphasized a rational planning model, his friend Alfred Smith cautioned him to realize just how great a force inertia exerted and pointed out that the general public required much more than just a careful, reasoned argument: "To get the people to see, you have, in the poetically expressed thought of the day, 'got to go after them with an axe, or a bludgeon.' The epidermis is so thick, and the general intelligence so dense—speaking generally—that you have to precede your effort by something resembling a mild earthquake. . . . Even then it is almost a hopeless task."[10] On the one hand, Wilgus could expect the same kind of opposition to collective planning from private business interests as he had encountered from the New York Central Railroad in plans for both the small-car freight subway and the Inter-Terminal Belt Line. And on the other hand, Smith warned of public apathy unless some crisis awoke the public to the cost of inaction.

Not to be deterred, Wilgus contacted New Jersey state senator James Johnson and presented his idea for an interstate transportation commission. Meanwhile in Albany, Alfred Smith suggested he could help with the New York legislature if Wilgus could convince Johnson to introduce a bill to form an interstate commission. In February, Wilgus sent Johnson his draft for a resolution to be introduced in the New Jersey Senate:

WHEREAS, Many autonomous communities in both the states of New Jersey and New York are embraced within the confines of the Port of New York;

WHEREAS, Each of such communities exercises local control of its waterfront facilities without cooperation, for the common good with other communities within the Port.

WHEREAS, . . . harmonizing of the relations of water and rail carriers . . . [and] the adoption of modern methods to the trans-shipment of freight between rail and water . . . urgently call for joint action, if the Port is to hold its preeminence among the ports of the world.

WHEREAS, An adequate study of the needs of the Port should demonstrate what action is desirable on the part of a public authority to bring about needed reform and improvements.

Now, THEREFORE, BE IT RESOLVED, That the Governor of [the] State of New Jersey is hereby authorized and requested to appoint three commissioners of experience and skill in matters relating to the construction and operation of port facilities for rail and water, to act jointly with three commissioners similarly appointed by the Governor of the State of New York.[11]

On February 15, 1911, Senator Johnson introduced Senate Joint Resolution no. 2 in the New Jersey Senate, copying Wilgus's text almost word for word.

The resolution reflected Wilgus's belief that a problem could be solved if a panel of experts carefully studied the problem and then recommended a course of action for the "common good." Proper action to solve the port problems would not result from the normal political process; only an "authority" of appointed, not elected, experts had the knowledge to see the larger picture and then would have the power to implement change. In return for giving up a degree of local control, all of the local communities would benefit in the long run. With power delegated from state government, the bi-state authority would stand above and beyond the parochial interests of local elected officials and state legislatures.

Although Wilgus succeeded in having his resolution introduced in the state senate in New Jersey, he did not achieve similar success in Albany. In the short run his plans for a bi-state port authority remained just another of his many ideas.

The Greater New York Belt Line

Efforts to develop a comprehensive plan for the port of New York had to address the need to improve both shipping and rail facilities. As we have seen, eight of the railroads serving the port terminated on the New Jersey side of the harbor. If not complex enough, interchanging freight entailed high costs and delays. The rail system resembled a "spoke and hub" system with multiple spokes running from outside the New York metropolitan area to the hub rail yards lining the New Jersey shore. Freight trains traveled from throughout the country along one of the spokes and delivered freight to rail yards for transfer to Manhattan or Brooklyn by car floats or lighters. Each railroad maintained a separate freight delivery system; any interchange of freight between railroads involved a cumbersome and expensive transfer process.

In a number of other cities, the railroads had banded together to build a belt line railroad at some distance from the city "hub," connecting all of the railroads serving the city. Freight cars could be interchanged over the belt line, dramatically increasing the efficiency of the overall freight system and reducing costs. Wilgus had worked for the city of Philadelphia when the city negotiated with the railroads to establish a belt line railroad.

An article in *Railway Age Gazette* in 1913 analyzed the inherent weakness in the freight system in the greater New York region. The heart of the problem lay in the fact that "New York has done less to develop its port along modern lines than any other important shipping centers in the world."[12] The article discussed a number of recent proposals to solve the problems, including one put forth in early 1910 by Calvin Tomkins, former commissioner of docks for New York City. Tomkins proposed construction of an elevated freight railway along the Hudson River in Manhattan connected to joint freight terminals and warehouses, and freight tunnels under the Hudson to a large rail yard in New Jersey for joint use by all of the railroads. His plan incorporated elements of both Wilgus's small-car freight subway and the Inter-Terminal elevated railroad on Manhattan's west side.

Wilgus carefully followed developments and strongly criticized the Tomkins plan.[13] In addition, he prepared a competing plan for an outer belt railroad of just over 60 miles in New Jersey from the West Shore Railroad to the north to Staten Island on the south. The outer belt line would cross over Arthur Kill by bridge and then across Staten Island to a tunnel under the Narrows connecting to Brooklyn. The tunnel under the Narrows between Staten Island and Brooklyn, whose cost would be shared with the Brooklyn Rapid Transit Company, would open direct rail access to the Brooklyn waterfront. In Brooklyn the planned belt line connected to the Long Island Railroad and the New York Connecting Railroad, providing "an all-rail route for New England traffic for all eight New Jersey roads."[14]

As with all his plans, Wilgus prepared a detailed cost model with two parts. First, he estimated the added costs to the New Jersey–based railroads of using the present system of moving freight by water across the Hudson River to Manhattan and Brooklyn. He calculated that it currently cost $7.65 million to move 18 million tons of freight. By comparison, moving the same tonnage via an outer belt line would cost an estimated $3.8 million, almost $4 million less. Building the outer belt, the second part of the cost model, would cost approximately $30 million, including interest and principal payments of $2 million a year. Net savings over and above expenses and interest for the outer belt would be over $2 million a year collectively for all of the railroads.

Wilgus's cost model assumed that all eight railroads would participate by advancing the necessary capital in anticipation of future savings. Realizing the advantages of the outer belt railroad required a long-term vision and the ability to project savings from the new freight system into the future. Far-sightedness and the ability to project into the future were qualities in short supply at the na-

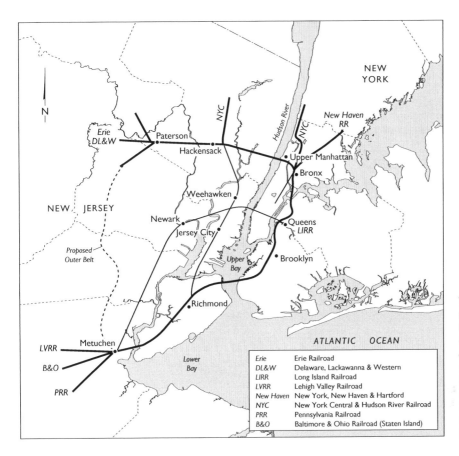

Wilgus's proposed outer belt line, 1913, connecting New Jersey–based railroads to Staten Island, Brooklyn, Long Island, the Bronx, Connecticut, and New England.

Map labels and legend:

NEW YORK

NEW JERSEY

Hudson River

NYC

New Haven RR

Erie DL&W — Paterson

Hackensack

Upper Manhattan

Bronx

Weehawken

Newark

Jersey City

Queens LIRR

Upper Bay

Brooklyn

Proposed Outer Belt

Richmond

ATLANTIC OCEAN

LVRR — Metuchen

B&O

Lower Bay

PRR

Erie	Erie Railroad
DL&W	Delaware, Lackawanna & Western
LIRR	Long Island Railroad
LVRR	Lehigh Valley Railroad
New Haven	New York, New Haven & Hartford
NYC	New York Central & Hudson River Railroad
PRR	Pennsylvania Railroad
B&O	Baltimore & Ohio Railroad (Staten Island)

tion's railroads. When Wilgus convinced the New York Central to proceed with the Grand Central Terminal project, he assured senior management and the board of directors that future revenues from air rights would more than cover the enormous sums borrowed for the construction. In this case, Wilgus's rational planning model with detailed cost estimates and future revenue projections carried the day. In fact, the Central did not realize the projected revenue from its real estate developments for decades to come.

Once again Wilgus believed that the inherent logic of the argument for an outer belt railroad and the cost savings to be achieved would convince all of the New Jersey railroads to participate. The model for organizing and financing the outer belt assumed that all of the railroads would jointly advance start-up capital and then form a corporation to run the outer belt railroad, with each railroad sharing stock in the new road. He described the plan as "a brilliant opportunity for reducing costs and improving the service of the railroads that serve the Port of New York."[15] He premised the plan on all of the railroads' benefiting from greater efficiency and lower costs over the long run.

Convinced of the soundness of the proposal, Wilgus went so far as to draw up a contract between himself and each of the participating railroads he thought would sign on immediately. The contract treated the outer belt proposal as his intellectual property, "whereas, THE ENGINEER [Wilgus] has brought to the attention of THE RAILROADS a preliminary study of an outer belt line in the vicinity of the port of New York."[16] The agreement then specified that if the railroads, or a portion of them, organized a company to finance, build, and operate the belt line, Wilgus would be entitled to compensation. The agreement specified that the "engineer" would receive 1 percent of the "estimated costs of constructing said outer belt line." If the outer belt cost the estimated $30 million, Wilgus's compensation would total $300,000, a fortune in 1913. To be fair, the contract called for Wilgus to develop detailed engineering plans for the outer belt and not just the eleven-page preliminary proposal.

Wilgus set out first to sell his idea for the outer belt railroad to the New Jersey–based railroads serving the port of New York. In October 1913 he wrote a detailed letter to E. R. Thomas, president of the Lehigh Valley Railroad. He included an analysis of the cost incurred by the Lehigh Valley of transferring freight, destined for New England, from the New Jersey waterfront to the New Haven Railroad in the Bronx at $875,000 a year. Exchanging freight with the Long Island Railroad by car floats across the harbor to Bay Ridge in Brooklyn added another $120,000. Additional costs arose from floating freight to the Brooklyn waterfront. Thus, moving 2,350,000 tons of freight cost the Lehigh a total $1,220,000. Using the outer belt as an alternative would cost $932,500 a year, with savings of $327,500 a year. He politely ended the letter by asking Thomas to respond "if in your opinion this preliminary study is sufficiently attractive to justify a further and more detailed investigation with a view to interesting all of the New Jersey roads." Thomas responded that the Lehigh would be very interested if the other railroads agreed to participate.[17]

In addition to the Lehigh Valley, Wilgus sought to convince the powerful Pennsylvania Railroad of the merits of the outer belt proposal. He wrote directly to the president of the Pennsylvania, Samuel Rae, and presented the same argument he had presented to the Lehigh Valley. The Pennsylvania would save over $600,000 a year and also gain revenue from the New York Connecting Railroad's increased freight traffic.[18] In total he projected savings and new revenue of more than $1 million a year. Wilgus added that he already had the support of President Thomas of the Lehigh Valley and offered to undertake a more detailed study, at no cost to the railroads unless construction of the outer belt moved forward.

After about a week Wilgus heard back from Rae and the Pennsylvania Railroad. Rae raised an immediate objection that the planned outer belt was too far from the harbor to attract the railroads. He suggested, as an alternative, improved rail access for freight to Manhattan by "several tunnels or one large bridge."[19] Of course, the Pennsylvania wanted freight access to Manhattan so it could compete directly with its archrival, the New York Central, which still enjoyed its monopoly on direct freight service to the island. Rae simply could not see beyond the need to compete with the Central. More efficiency for all of the New Jersey railroads by using the outer belt remained too abstract a proposition. Not for the last time in his career Wilgus ran straight up against the obstinacy of the private railroad companies. In the short run the Pennsylvania could not see an immediate payoff for participating in the financing of an outer belt railroad. Absent any pressure from political forces in either New Jersey or New York, the railroads seemed content to continue freight operations as they always had no matter what cost or complexity was involved. Since the outer belt line plan served the entire bi-state port of New York, neither the state of New Jersey nor the state of New York—never mind New York City or any of the other local municipalities in the region—held the power to force the railroads to participate in the belt line or any other comprehensive plan.

Back to Port Authority Planning

Public pressure to solve the problems of the port of New York continued over the next few years. Wilgus and other engineers and transportation advocates increasingly feared that the chaos in the port and high costs threatened New York's supremacy as the country's leading port. Baltimore, Boston, Philadelphia, and even Portland, Maine, were not sitting idly by. Shipping companies and railroads operating in these ports worked hard to capture some of New York's enormous freight traffic, including the city's major share of the country's imports and exports.

In December 1916, Wilgus wrote a long letter to Henry Hodge, chairman of the Public Service Commission, the New York state agency charged with overseeing transportation and utility matters in New York City. Once again he returned to the chaotic, inefficient freight transportation in the port. He emphasized the key problem: the lack of any central authority and the total unwillingness of the private railroad and shipping companies to cooperate in any substantive way. With World War I already raging in Europe and the possibility of U.S. involvement increasing, Wilgus added a national defense argument for solving the port's problems. He pointed Hodge to the recent formation of the Port of London Authority as a model and emphasized not only that "the time is ripe for doing this," but also, "above all," it was necessary for "perfecting our means of defense."[20]

Beyond national defense, the continued internecine warfare between shipping and rail interests in New York and New Jersey threatened to undermine the city of New York's dominant commercial position. For years, the New Jersey–based railroads and shipping companies had objected to the system of charging one freight rate for the entire port. The Interstate Commerce Commission essentially controlled freight rates the railroads charged across the country. New Jersey interests complained bitterly to the ICC that the railroads should charge a lower rate for freight delivered to the New Jersey shore of the port than to the piers in Manhattan and Brooklyn. The railroads viewed the rate structure as unfair because they had to absorb the costs of moving freight across the Hudson River.

In May 1917, Jersey City, Hoboken, and Newark, later joined by the state of New Jersey, filed a formal complaint with the ICC arguing that the agency should establish lower rates for freight delivered to the New Jersey side of the harbor.[21] If the railroads then moved the freight across the harbor, they would have to charge an additional fee. Officials in Newark and Jersey City and in the state capital in Trenton believed that with a "freight differential," a great deal of the port's shipping business would move across the harbor to the New Jersey side.

Wilgus, among others, recognized the seriousness of the challenge posed by the *New York Harbor* case before the ICC. In a private letter to Henry Hodge of the PSC, he warned that the imposition of a freight differential would threaten the fabric of New York's commercial dominance: "The pending proceedings before the Interstate Commerce Commission that have been brought by the New Jersey communities . . . , if approved by the Commission, will seriously endanger some eight billion dollars of property on the New York side of the Port." The "eight billion dollars" referred to the enormous investment in the transportation infrastructure in the city. He returned to the theme of comprehensive planning for the port: "I do not see how the various communities embraced within the limits of the Port can continue to dwell together . . . unless all differences can be reconciled and all pull together on some predetermined plan to be mapped out as I have attempted to explain."[22]

Wilgus followed this letter, which urged the PSC to establish a joint commission to study the entire port of New York, with a letter to the governor of New York, Charles S. Whitman. Here, Wilgus described himself as motivated "solely by a desire to promote, in the public interest, an impartial and *effective* joint inquiry by the States of New York and New Jersey and the National Government into the urgent needs of the Port of New York." He warned that "a failure of these two states and of the Nation now to take up this problem with the intention, in good faith, of overcoming legal and other obstacles and successfully solving it, will spell disaster for the future."[23] He offered to meet the governor either in New York City or in Albany to discuss his ideas further. Wilgus was in fact not simply a disinterested bystander. In previous years, he had worked hard to secure freight and passenger franchises and most recently had promoted his plan for an outer belt railroad.

Efforts to find a solution to the port's problems moved forward on several fronts. In March 1917 the press reported that both Governor Whitman and Governor Edge of New Jersey endorsed the plan for a bi-state study commission.

The efforts of New York interests to counteract the demands for rate differentials in the *New York Harbor* case before the ICC involved the New York Chamber of Commerce and the chamber's lead counsel, Julius H. Cohen. Cohen led the legal fight before the ICC, arguing that historically the port of New York had always been treated as a single district. At the same time, the chamber of commerce actively pushed to form a bi-state agency to study the problems of the port—just as Wilgus advocated for years. If the study resulted in a dramatic improvement in the overall efficiency of the port, rate differentials would not be necessary; the New Jersey side of the harbor would prosper along with the entire port. Jameson Doig, in his study of the Port Authority, also notes that if the "governors could be persuaded to take the lead in pressing for cooperative action, the ICC might be willing to hold the litigation in abeyance while the two states tried to reconcile their differences."[24]

Political maneuvering continued, and in March both governors publicly en-

dorsed the idea of a bi-state study commission. A few weeks later the state legislatures passed the necessary bills to set up the New York, New Jersey Port and
Harbor Development Commission.[25] William H. Willcox, former chairman of
the PSC in New York City, chaired the new commission; and Eugene Outerbridge, a Staten Island resident and longtime transportation advocate, joined
Willcox and George Goethals of Panama Canal fame. The ever-present Julius
Cohen joined the commission as chief counsel.

Wilgus had argued for almost a decade for the establishment of a bi-state study
commission as the essential first step in solving the port's problems. By now,
however, he was in Europe with the Allied Expeditionary Forces organizing a
transportation network in France to supply the American forces and could not
serve on the commission he had worked for so diligently. Histories of the Port
Authority start with the formation of the Harbor Development Commission and
the work of Julius Cohen. History proves to be complex, and major developments,
such as the rise of the Port Authority, often have more than one author. Prominent
among them, in the case of the Port Authority, stands William J. Wilgus.

Transportation in the Great War

Like most Americans, Wilgus followed world events with concern as Europe
drifted toward war. He described his feelings after the Great War broke out in
the summer of 1914: "My sympathies, like those of a substantial proportion of my
countrymen, were on the sides of the Allies in their struggle with the Central
powers in Europe. As the submarine sinkings at the hands of the Germans grew
both in number and horror, in which American lives and ships increasingly lost,
I joined a wave of indignation . . . by marching up Fifth Avenue."[26] The sinking
of the *Lusitania* on May 7, 1915, off the coast of Ireland, evoked enormous outrage, including the massive protest march up 5th Avenue, which Wilgus joined.

A year later, Wilgus received a letter from the War Department in Washington asking that he volunteer for service in the army. As America prepared to join
the conflict, the War Department knew that an adequate supply system would
be crucial to support an American army in Europe. A supply system would be
totally dependent on rail transportation, and the army needed experienced railroad engineers. Wilgus did not hesitate: "I could but pause a moment before
deciding on the course I should pursue. To sign and transmit the completed form
would mean a sharp break in my chosen career of independence at the peak of
my earning power. . . . I reached for the pen and signed my name."[27] In October
of 1916, at the age of fifty-one, he received a commission as a colonel in the Railroad Transportation Corps. For the rest of his life, friends, business associates
and the press often referred to him as "Colonel" Wilgus, a title he encouraged.

Wilgus entered active service in May of 1917, just weeks after the United
States had declared war on April 6, 1917. As a member of the newly appointed
Military Railway Commission, he set sail for England and France to "ascertain
the transportation needs of the Allies."[28] A daunting task awaited the commission. With the French and British having already fought the Germans in France

and Belgium for almost three years, the French rail system had been stretched to its limits. The commission immediately realized that when American forces began to arrive in large numbers, they would have to organize and operate a separate transportation system to deliver supplies to the front for their own troops. The commission immediately got to work to study existing facilities and plan for needed railroad equipment and personnel to be requisitioned from the United States.

The experience that Wilgus had gained as a young railroad engineer equipped him to help with the planning task in France. When he had first gone to work for the New York Central Railroad, he had undertaken detailed field observations about the state of the railroad's subsidiary, the Rome, Watertown Railroad, and had drawn up plans to rehabilitate the road. In a similar fashion in France, the members of the Military Railway Commission, after seventeen days of nonstop night and day work, completed a report and forwarded their recommendations to the chief of engineers in Washington. While the work in France was on a much larger and more complex scale than Wilgus had undertaken in New York, the essential analytical work—careful data collection, organization, analysis, and synthesis—remained the same. Among the crucial estimates to be made were the size of the army; the number of ports and rail routes needed; and the numbers of experienced American railroad workers required to run ports, maintain the track system, and operate the supply trains.

Wilgus, on behalf of the commission, wrote a detailed letter to the commanding general of the lines of communication in which he concisely, as always, laid out the scope of the problem and the needed port and rail facilities. He recommended the construction of two or more ports, maintained and operated by American engineers, that would be capable of handling 40,000 to 50,000 tons per day in addition to troops. Wilgus and the commission assumed that the American forces would total 500,000 to 800,000 troops, and their requests for railroad equipment and trained personnel reflected this estimate. These initial estimates required two or more lines of communication between different ports and American forces on the front.[29]

Creating a Transportation Organization

After the commission's report was sent to Washington, Wilgus remained in France. In August he received a new title, director of military railways, with an office in Paris at 149 Boulevard Haussmann. He reported directly to General John J. Pershing, commander of the American Expeditionary Forces (AEF) in Europe. Working nonstop, Wilgus and his staff drew up the detailed plans to set up the army's Transportation Service in France. Significant responsibility fell on his shoulders: "In all fairness," he said, "I was to be held responsible for the success or failure of the outcome."[30]

Thus, for the second time in his career, Wilgus took the lead in a massive undertaking. In 1902 in New York, as fifth vice president of the second-largest railroad in the world, he had set out to rebuild a major section of midtown Manhattan, anchored by the magnificent new Grand Central Terminal. Sixteen years later, he was planning and directing the transport and supply of more than two million American soldiers to Europe and on to the front. Never hesitant to assume responsibility and never doubtful of his own abilities, he simply got on with the task.

From their very first meetings, General Pershing developed great trust in Wilgus's abilities and endorsed his first reports detailing existing transportation conditions in France and projecting equipment and personnel needs. Wilgus reported, "I was told by him that he approved my recommendations in their entirety, that I should report to him directly as required by circumstances regardless of rank, that my headquarters might be established wherever I chose."[31]

Wilgus's plans called for an integrated transportation system that extended from the ships arriving at French ports to the troops on the front lines. One command, with the power to eliminate bottlenecks and move supplies efficiently, would exercise total responsibility. Above all else, Wilgus feared a divided effort, with numerous army departments, each with a limited area of responsibility, constantly squabbling and creating bottlenecks. Just as he had sought to bring order and integration to the chaotic transportation system in the port of New York, with rail transportation divided among numerous competing railroads, in Europe he set out to build an integrated logistics system to supply the AEF by rail.

Organizing the Transportation Infrastructure

In the spring of 1918, the deteriorating military situation and the real fear of a German breakthrough brought desperate calls from the American allies to increase the number of troops destined for Europe. The Transportation Service received orders to prepare to supply an army numbering over 1.4 million men by September, a 50 percent increase over previous estimates; and the number of troops was expected to total 2 million men by December 1. The projected increases in troop strength dramatically expanded the task before the Transportation Service, as estimated supply needs for this number would exceed 50,000 tons a day (table 5.2).

In his later account of these years, Wilgus describes the efforts undertaken to accommodate this rapid buildup of troops. Encyclopedic in nature, *Transporting the A.E.F. in Western Europe* makes clear just how complex the assignment was—as do the careful records that Wilgus kept of all his

Table 5.2. Projected increases in U.S. troop strength and needed supplies for July 1918 through July 1919, European theater, World War I

	No. of troops	Avg. daily tonnage
1918		
July 1	1,115,486	28,000
Sept. 15	1,441,336	36,000
Oct. 1	1,616,336	40,000
Nov. 1	1,866,336	47,000
Dec. 1	2,091,336	52,000
1919		
Jan. 1	2,291,336	57,000
Feb. 1	2,491,336	62,000
Mar. 1	2,741,336	69,000
Apr. 1	3,041,336	76,000
May 1	3,376,336	84,000
June 1	3,726,336	93,000
July 1	4,026,336	101,000

SOURCE: Wilgus, *Transporting the A.E.F. in Western Europe, 1917–1919* (New York: Columbia University Press: 1931), 211.

activities while overseas. These materials, which fill eighteen boxes in the manuscript collection of the New York Public Library, reflect his ability to assemble and analyze data and then to devise an efficient solution. No detail escaped attention, from the number of berths needed at ports throughout France to the number of cranes needed to lift cargo to "miscellaneous tools and appliances" needed at railroad repair shops.[32] Wilgus believed the work of the Transportation Service played an absolutely crucial role in the outcome of World War I. In his memoirs he observed: "A failure to deliver men, ammunition, guns or subsistence at the crucial moment may mean all the difference between victory and defeat."[33]

To supply the expanding number of AEF troops on the western front, the Transportation Service established a number of separate lines of communication to transport supplies from a port to the front. Each line of communication started at a port with numbers of berths for ocean-going ships and unloading facilities, including warehouses for sorting of incoming supplies. Railroad cars

France, World War I, 1918: fifth line of communication for the American Expeditionary Forces (AEF) from Brest to the western front.

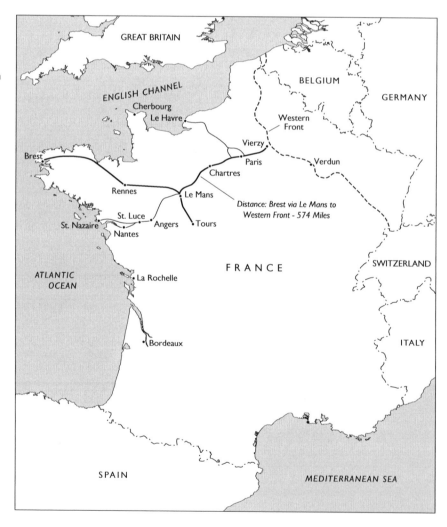

Table 5.3. The fifth line of communication, Brest to Vierzy, November 1917

| From | To | Miles | % grade | Assumed train load[a] | | |
				Gross (tons)	No. of cars	Net (tons)
Brest	Le Rody	4	1.12	1,050	29	525
Le Rody	Rennes	151	1.12	1,050	31	563
Rennes	Le Mans	100	0.68	1,640	49	880
Le Mans	Grievres	118	0.72	1,570	45	809
Le Mans	Chartes	77		1,330	40	720
Chartes	Trilport	92		1,630	45	807
Trilport	Vierzy	32			30	

SOURCE: Wilgus, *Transporting the A.E.F.*, 240.
[a] Gross weight = weight of engine, cars, and freight; net weight = weight of freight only.

were loaded, assembled into trains, and then hauled across France to a staging area near the front. From the staging area to the front lines, the army used narrow-gauge railroads to move supplies as close to the front as possible.

The first line of communication ran from three primary ports on the Atlantic: St. Nazaire, La Rochelle, and Bordeaux across to the western front. The total distance varied between 600 and 650 miles, with numerous assembly points in between, and a total of 1,646 miles of main line tracks. Soon second, third and fourth lines of communication needed to be added. By November 1917, the Americans needed a fifth line and utilized Brest in Brittany as the port to supply the western front at Vierzy, 574 miles away (table 5.3).

Ports and Operating Equipment

Based on his first inspection tour, Wilgus knew that establishing an efficient supply system depended on adequate port facilities. The port situation created major difficulties for the AEF. As Wilgus recalled in his book, "No problem of the war would press upon us more heavily than the one that had to do with the ports of France. . . . The gateways to France should be located within the shortest possible distance from the army's objective, and they should be adequate in number and efficient in purpose."[34] The only French ports available, on the Atlantic Coast, did not meet the army's requirements: they had neither adequate deepwater berths nor sufficient railroad facilities alongside the piers.

In July 1917, Wilgus forwarded to Washington plans to modernize the existing ports by building new piers and rail yards, an effort that required additional men and material. These plans "called for extensive alterations and changes at existing ports and the construction of new ones on an imposing scale."[35] New, efficient port facilities would make possible a dramatic increase in the movement of supplies, especially with the addition of a modern railroad infrastructure.

Left: Constructing new pier facilities in France, 1917: American engineers and construction troops building new piers and a rail yard in a French port to supply the Allied Expeditionary Forces.

Below: Unloading supplies, France, 1917: American service troops and stevedores unloading supplies destined for the western front.

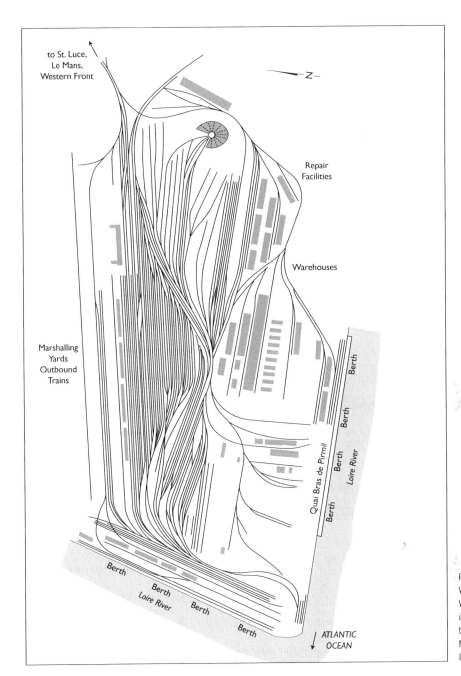

to St. Luce,
Le Mans,
Western Front

– Z –

Repair
Facilities

Warehouses

Marshalling
Yards
Outbound
Trains

Quai Bras de Pirmil

Berth
Berth
Berth
Berth
Berth
Loire River

Berth
Berth
Loire River
Berth
Berth

ATLANTIC
OCEAN

France, World
War I, 1918:
Wilgus's
improvements
to the port of
Nantes on the
Loire River.

Over time (much too slowly in Wilgus's view), the number of ports and piers for use by the Transportation Service increased. In addition, improved unloading facilities increased the tonnage of supplies off-loaded from ships each day and loaded onto waiting freight trains. By June 1918 these facilities had reached a daily capacity of 101,000 tons (table 5.4)—a remarkable achievement, given the existing conditions when the first troops began to arrive in France in large numbers.

Table 5.4. Port capacity in France, June 1918

	Berths	Coal	Freight	Total
		Est. daily port tonnage		
Siene Group				
Le Harve	8	1,000	2,500	3,500
Rouen	4	1,500	2,500	4,000
Upper Coast Group				
Caen	2	1,000		1,000
Cherbourg	2	1,700		1,700
Saint-Malo	1	2,000		2,000
Brest	7	1,600	6,000	7,600
Loire River Group				
Saint-Nazarie	15	1,500	7,500	9,000
Montoir	8	1,500	3,000	4,500
Donges (ammunition)	4		3,300	3,300
Nantes	19	2,600	4,400	7,000
La Pallice	6		5,000	5,000
Rochefort	6	1,500	2,500	4,000
Gironde River Group				
Pauillac	6		2,000	2,000
American Bassens	10		11,000	11,000
French Bassens	10	1,500	3,500	5,000
Bayonne	4	1,900	600	2,500
Mediterranean Group				
Marseille	14		10,000	10,000
Toulon (ammunition)	3		2,500	2,500
Total	150	30,000	71,000	101,000

SOURCE: Wilgus, *Transporting the A.E.F.*, 244.
NOTE: Minor ports are not included.

Securing adequate port facilities constituted a first step in the supply chain. The Transportation Service also set up its own operating division to run supply trains to the front. Wilgus also knew from the very beginning of planning that to supply the AEF, the Americans would have to equip and operate their own trains with railroad workers brought over from the United States. The French rail system simply did not have any excess capacity.

The needed locomotives and cars had to be ordered as quickly as possible to ensure adequate rolling stock. Wilgus's initial plans asked for 3,555 locomotives and 90,960 freight cars, numbers regarded in Washington as "excessive." To provide a comparative perspective, in 1916 the Erie Railroad, one of the major trunk line railroads operating in the eastern United States and servicing the port of New York, operated with a total of 1,461 locomotives and 51,735 freight cars and hauled 44,359,314 tons of freight.[36] Wilgus and the Transportation Service planned for moving over 100,000 tons of supplies each day to the front, equivalent to 36.5 million tons of freight a year, almost as much as the entire freight traffic of the Erie Railroad in 1916. In a real sense, Wilgus and the Transportation Service were being asked to create a railroad system in France with the freight capacity of the Erie Railroad in just a few months; to have it up and running as hundreds of thousands of American troops arrived and quickly moved to the western front; and then to provide the AEF with all needed supplies.

Despite the best of plans, the sheer magnitude of creating an operating railroad system of the size and scale required proved to be a daunting task. Almost immediately, the Transportation Service fell behind in meeting the daily tonnage needs of the AEF. No matter how heroic the efforts of the Transportation Service, shortages persisted and loud complaints filtered up the chain of command. Wilgus identified many reasons for the shortages, including slow progress in the construction of new port facilities and an inadequate number of trained railroad workers to operate the supply trains.[37] Despite all of the problems, the achievements of the Transportation Service in the brief period of time between June 1917 and the armistice in November 1918, summarized in table 5.5, were impressive.

Wilgus and all who served in the Transportation Service could justifiably be proud of their accomplishments. In just eighteen months, almost two million American soldiers were transported to France and supplied with 10 million tons

American supply and troop trains arriving at the western front, 1918.

Table 5.5. Accomplishments of the Army Transportation Service, June 1917 to May 1919

| | Cargo shipped (tons) | Animals shipped | No. of troops transported | |
			To France	Back to U.S.
1917				
June–October	299,453	2,899	90,356	
November	124,641		29,521	
December	113,825	3,774	37,552	
1918				
January	162,016	3,694	47,763	
February	192,239	6,559	33,455	
March	288,038	8,547	62,348	
April	397,969	2,232	84,948	
May	478,197		162,885	
June	607,274		221,288	
July	641,959		301,526	
August	715,258		242,207	
September	767,648	1,839	311,969	
October	919,488	2,582	217,614	22,183
November	920,972	20,881	112,626	25,503
December	910,059	11,801	6,574	98,669
1919				
January–May	2,120,263		5,025	1,128,896
Total	9,577,943	64,918	1,967,627	1,275,251

SOURCE: Wilgus, *Transporting the A.E.F.*, 444.

of supplies. In Wilgus's summary, "In essence, [our] plans called for the completed establishment in short order on foreign soil, 3,000 to 5,000 miles from home, of means for the operation of ports and lines of communication that rivaled in size and importance the greatest railway . . . systems in the United States . . . and the putting of them into successful operation with quickly assembled personnel."[38]

Just after the armistice, Wilgus decided to resign. In his formal resignation letter to Pershing, dated November 17, 1918, Wilgus requested permission to return to the United States because of "the urgent need, for personal reasons, of my presence at home." Wilgus followed with a confidential letter directly to Pershing: "I have given to the Army the very best that is within me. . . . I am very anxious to return to the United States, where the recent loss of a close member of my family makes my presence highly desirable."[39]

The close member of his family referred to was his wife, May Reed Wilgus, who had died suddenly at the Hotel Seville in New York on October 2, 1918. A very private man about his personal life, Wilgus could not bring himself to explain to Pershing that his wife had passed away the previous month. Except for one trip back to the United States, Wilgus had served in France, apart from his wife, for over a year and a half and did not even return to the United States for her funeral.

Pershing replied to Wilgus on December 2, 1918, informing him he would do everything in his power to speed up approval of his resignation. In this letter Pershing praised Wilgus's work in France:

> I cannot overlook this opportunity to let you know briefly how much your services in the Transportation Service have been appreciated. . . . Those associated with you, especially myself, have had the fullest confidence in you and have always regarded your clear and comprehensive plans as our guide. To you, more than any other man is due the credit for the great construction program carried out. . . . Most pleasantly shall I retain in my memory the high example of that devotion so unselfishly given, which has at all times characterized your work. It has been a distinct pleasure to know you personally, as well as officially, and I shall hope that our friendship may continue in years to come.[40]

High praise indeed from the commander of the American Expeditionary Forces, known as a hard man with subordinates whose performance proved to be less than satisfactory to his high standards. Pershing never suffered fools gladly. General Pershing followed his warm letter with a recommendation to Washington and the president that Wilgus be awarded the Distinguished Service Medal. A letter to Wilgus dated March 12, 1919, informed him of the award; more honors followed, including the Légion d'Honneur from the French government on April 4, 1919.[41]

In his memoirs, Wilgus concluded his narrative of his work in Europe by simply stating, "The need for my services now having ended, I tendered my resignation to General Pershing . . . said goodbye to my beloved fellow officers and associates and left by the steamer *Espange* for New York."[42] He arrived home in New York on January 1, 1919. After almost two years in France, he faced beginning anew his consulting career at the age of fifty-three, not an easy task, as he explained: "The problem of making a new start in life . . . faced me when I bade farewell to the Army and again hung out my shingle. My old clients, of necessity, had turned meanwhile to other consultants for advice, and my vacant office at 165 Broadway, New York had gathered dust while I was so long away."[43] His partner in the Amsterdam Corporation, Henry Pierce, had already left New York for Seattle to become president of the local street railway company. The Inter-Terminal Belt Line syndicate, formed to promote the elevated railroad and subway circling Manhattan Island, quietly dissolved. Once again he immediately turned his considerable energies and expertise to solving the transportation problems in the port of New York.

A Unified Port of New York

Wilgus found New York City in the midst of a continuing revolution in transportation. With the dual system expansion of the subway well under way, the lengthened subway had stimulated growth in the outer boroughs. Simultaneously, the number of automobiles and trucks on the roads had grown exponentially; and in both New York City and the wider metropolitan region, the streets and avenues had become increasingly crowded with vehicular traffic. Political pressure on the Public Service Commission to find solutions had risen, and in 1919, New York State merged the first and second districts of the PSC and then in 1921 replaced the PSC with the New York State Transit Commission, chaired by George McAneny. In the period just after the war, McAneny and the Transit Commission focused all of their energy on the city subways and not on the larger question of improving transportation throughout the port of New York.[44]

Wilgus's experiences in Europe had reinforced his conviction that solving the problems of the port of New York required an integrated system. No one railroad or shipping company could solve the challenges confronting the busiest port and manufacturing complex in the United States. In Europe, the army's Transportation Service had the overall responsibility for transporting and supplying the AEF. One command oversaw the entire supply system from the ports on the Atlantic Ocean to the American soldiers on the front lines. When problems inevitably occurred, it fell to the Transportation Service to work out a solution, with the overriding goal of maintaining a multifaceted transportation system from the multiple ocean piers, to the five main lines of communication (in reality, a series of railroads crisscrossing France), to warehouses and supply depots near the front. Authority rested in one place, and along with concentrated power came great accountability. As Wilgus dramatically argued in his book, an integrated supply system contributed to victory.

During the war the expanded New York subway system had improved internal circulation of passengers in the city and the outer boroughs, but the overall transportation system in the port remained chaotic. Freight traffic to the port had increased dramatically. By 1917 the port's inherent inefficiency had led to an almost complete breakdown in moving freight. Thousands of tons of freight waited on the piers throughout the harbor to be loaded onto ships bound for France. Because of the huge bottleneck in the port, freight trains destined for New York could not move and were sidetracked and delayed all the way back to Chicago. With freight sitting on sidings, a severe shortage of freight cars developed, and shippers could not load freight at factories or warehouses. As much as the railroads promised to improve efficiency, they remained hampered by their inherent competitiveness.

Legislation passed by Congress in August 1916 in the run-up to the war authorized the president to take control of the nation's railroads in order to move troops and supplies if congestion and delays threatened the war effort. As soon as the United States entered the war, railroad executives met in Washington and pledged to cooperate in an effort to avoid an outright government takeover.[45] Yet the transportation crisis continued, and in December 1917, President Wilson decided to act: he nationalized the railroads and appointed William McAdoo, builder of the Hudson & Manhattan Railroad, as director-general of the U.S. Railroad Administration. With direct power over all railroads serving the port of New York, the Railroad Administration quickly brought some order to the chaotic freight handing and shipping in the harbor.

The key to the Railroad Administration's success in the case of the port of New York rested on its power to organize and control the entire harbor. Ships were assigned to piers where their cargo awaited, regardless of which private company or railroad owned the pier or the ship. Increasing numbers of ships were loaded on the New Jersey side of the harbor, much closer to the rail yards of ten of the twelve railroads serving the port, avoiding the time-consuming and inefficient transfer of freight across the Hudson. Given the wartime emergency, the Railroad Administration was employing a rational planning model in the port of New York and exercising direct control over all of the railroads in both New York and New Jersey.

When World War I ended, the work of the Railroad Administration continued for another year until it officially ceased to exist in March 1920 and all of the railroads returned to private management. The lessons learned during the war encouraged advocates and civic leaders to call for creating a permanent agency to oversee the entire port. In March 1919, just a few months after the armistice, the Merchant's Association of New York had called for the creation of a port authority to plan and coordinate improvements throughout the two-state harbor. The Port and Harbor Development Commission, the bi-state agency established in 1917, just as Wilgus was beginning his years with the army, had also proposed creation of an agency with jurisdiction over the entire port. The Merchant's Association pointed to the danger of New York and New Jersey each operating in their own narrow self-interest and encouraged members to support "the pending Port Treaty between New York and New Jersey [which] provides for such joint plans and joint management."[46] A direct line of development runs from the New York, New Jersey Port and Harbor Development Commission (1917) to the Port of New York Authority (1921), today simply referred to as the Port Authority.

Given his long involvement in New York's transportation issues, Wilgus carefully followed all of the discussions and debates over the future of the port. His plans for a small-car freight subway to link New Jersey and Manhattan had certainly addressed the freight problem in the port in a comprehensive way that included linking the twelve railroads providing freight service to the harbor. Wilgus also claimed that the very idea of a bi-state port authority, with jurisdiction over the entire harbor, was his. Among his voluminous papers in the Manuscript Collection in the New York Public Library is a folder entitled "Movement Leading to N.Y.–N.J. Interstate Legislation Respecting a Port of New York Authority."[47] The cover memo, dated March 24, 1933, starts with a bold assertion: "The idea of a joint movement on the part of the two states of New York and New Jersey for the creation of a 'Metropolitan Freight Transportation District'

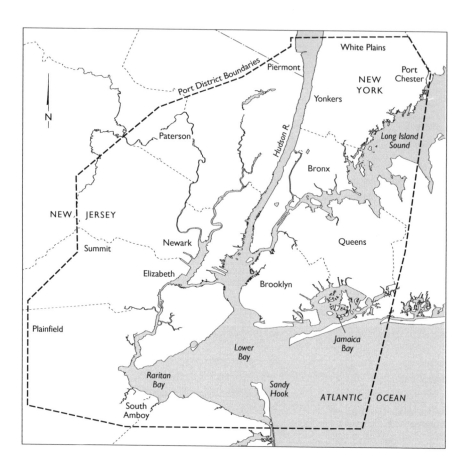

Wilgus's boundaries for the port of New York, 1911.

embracing all parts of the port, was first given public expression by the author [William Wilgus] in an article appearing in the Engineering News on March 31, 1910." Wilgus claimed that seven years before the Harbor Development Commission and a decade before the Port Authority, he had already thought through the need for a portwide authority to bring order to the transportation chaos in the New York metropolitan region.

In his history of the Port Authority, Doig includes reference to Wilgus and mentions a number of his plans for improved freight service, but he does not credit Wilgus as one of originators of the idea for a port authority. Instead, he credits Julius H. Cohen, a lawyer and public advocate, with first laying out the rationale for a bi-state port agency in an article in the *Cornell Law Quarterly* published in 1920.[48] Cohen served as counsel to the Harbor Development Commission and had also served as counsel to the New York Chamber of Commerce in the famous *New York Harbor* case, discussed earlier.

The New York, New Jersey Port and Harbor Development Commission

The Harbor Development Commission began its work in August 1917 and immediately started recruiting staff and planning for the major study of the port of New York. The commission requested $400,000 for a period of two years to complete the needed detailed data collection and analysis.

With the Harbor Development Commission's work under way, the ICC decided the *New York Harbor* case on December 17, 1917, declining to order the railroads to charge differential freight rates for each side of the harbor. In the decision the ICC found that "historically, geographically, and commercially New York and the industrial district in the northern part of the state of New Jersey constitute a single community."[49] But while the ICC declined to order rate differentials at the present time, it warned that if overall improvements to freight handling in the port were not made, it would revisit the issue and consider imposing rate differentials. The ruling from the ICC only added greater impetus to the work under way.

The Harbor Development Commission organized the in-depth study of the port into three broad categories of investigation: (1) a study of existing conditions; (2) an analysis of conditions in terms of output and costs; and (3) preparation of a comprehensive plan of development for the future. To carry out the actual work, the commission set up four divisions. An inspection force would make "clockings" of railroad, steamship, ferry and trucking operations and estimate the costs of new facilities. A "statistical force" would calculate freight tonnages and costs of present transportation operations. Specialized engineers would complete studies of market and food distribution, freight-handling machinery, and water supply. Draftsmen would make maps and drawings

All of the commission's work rested on a single underlying premise: that the port of New York encompassed a unified, bi-state metropolitan district. For working purposes the commission defined the geographical boundaries of the port district: Yonkers and New Rochelle on the north; Great Neck and Rockaway on Long Island to the east; Perth Amboy and Sandy Hook on the south; and Paterson, Newark, and New Brunswick on the west. This area was home to over six and a half million people in 1910; estimates projected the population to increase to over twenty million residents by 1950. At the core of the region, the port of New York constituted the greatest commercial and industrial center in the country, as the preamble to the commission's first report had detailed:

NEW YORK AS A PORT

New York is by far the greatest railroad, commercial and industrial center on the Atlantic seaboard.

The population of the metropolitan district is four times that of Philadelphia.

The Port is the finest harbor on the Atlantic Coast, with 780 miles of waterfront.

Three times as many railroad trunk lines come to New York as come to any other city on the Atlantic tidewater.

More manufacturing industries are located in the New York district than in the combined cities of Philadelphia, Chicago, Cleveland and St. Louis, and the value of products of the New York industries is nearly as great as the combined value of the products of those of the other four cities.

Approximately half of all foreign commerce of the United States passes through the Port of New York.[50]

Above all else, the commission focused on the two major components of the port's transportation system—railroads and shipping. The charge to the commission did not include any mention of the region's highways or vehicular traffic. In 1917, when the commission began work, the railroads, and the railroads alone, transported all of the vast volume of freight that moved to and from the port each year. In turn, the shipping companies carried freight to the entire world, along the coast, up the Hudson River, into Long Island Sound, and on to New England. Only a few two-lane highways linked the port of New York and the metropolitan region to the rest of the country. In 1917, America's internal transportation system remained totally dependent upon the private railroad companies. Beyond the shore, shipping companies provided all transportation. Aviation was in its infancy; the first commercial airline service from New York did not begin until June 1926. Transporting freight by air was far in the future.

In a first report the commission identified the railroads as the key problem:

> Unquestionably the most important physical problem of the Port is providing for better and more economical movement of inbound and outbound freight over the trunk lines now terminating on the New Jersey and Staten Island waterfront. . . . Three problems that are in a way part of the general problem . . . are the West Side railroad situation in Manhattan; the food distribution system, which is inseparably related to the West Side Problem; and the exterior belt line in New Jersey . . . Another major problem . . . is that of providing adequate steamship terminals properly related to the railroads.[51]

Wilgus had already proposed a solution for the more economical movement of freight with his small-car freight subway in 1907. For the west side problem, in 1910 he had proposed the Inter-Terminal Belt Line as a way to remove freight trains from west side streets. Further, in 1910 he had tried to convince the trunk line railroads to invest in a belt line railroad, a major portion of which would serve as the outer belt line in New Jersey, something that the Harbor Development Commission identified as a key part of any plan to improve freight transportation.

In 1920, after two years of work, the commission published its *Joint Report with Comprehensive Plan and Recommendations*. Divided into three parts, the 495–page report included voluminous data and analysis of present port conditions and presented a comprehensive plan for improvement. No part of the port and shipping operations escaped attention. One chapter covered the history of waterfront policy going back in time to the Dutch settlement. Another section of the report examined warehouse facilities throughout the port area; and another delved into the distribution of milk, dressed meat, fish, oysters, and potatoes. Needed shipping channel improvements were investigated along with the development of Jamaica Bay as a deepwater auxiliary port to handle international shipping. Transporting fuel, grain, and ice merited separate chapters, as did electric power, water supply, and municipal waste.

The work of the Harbor Development Commission represented the epitome of the rational planning model advocated for so long by Wilgus. Finally a bi-state commission had the resources and time to gather all necessary data, analyze the information, and develop recommendations for improvements. With far fewer resources Wilgus had utilized the same analytical model when developing plans for the small-car freight subway, the Inter-Terminal Belt Line, and the outer belt line railroad.

The Railroads

The *Joint Report* reiterated the crucial part the railroads played: "Our port problem is primarily a railroad problem. Navigable water, or water easily made navigable, reaches into every part of the Port District, while rail accessibility and co-ordination are lacking at many points. Therefore the comprehensive plan to evolve with this Commission as created is essentially a railroad plan. . . . A complete reorganization of the railroad terminal facility is the most fundamental physical need of the Port of New York."[52] Given proper planning and coordination, the commission report envisioned a more efficient freight transportation within the port based upon dramatic improvements in the railroad infrastructure. The report envisioned cooperation among all of the private railroads serving the port and the metropolitan region.

The sheer magnitude of freight handled in a single year illustrated the challenge the planners faced. In 1914 the railroads had hauled 76 million tons of freight. To try to put this amount into perspective, the *Joint Report* explained: "If all of this freight had been loaded simultaneously into cars it would have filled a track on each of the eight trunk lines across the continent."[53] During World War I, the army Transportation Service moved a total of 9,577,943 tons of freight to Europe between June 1917 and May 1919. These supplies sustained the American Expeditionary Force—over two million men—and also provided the other allies with needed materials. The amount of freight handled in the port in just one year dwarfed the achievement of the American supply services during World War I in which Wilgus had played such a prominent role.

To analyze freight movement by the railroads, the commission carried out an exhaustive data collection process. The research team decided that relying on available statistics would not be sufficient. At all of the major "break-up" yards where the railroads unloaded and loaded freight—waterfront rail yards, pier stations, and inland freight depots—teams conducted clockings, recording all movements of trains, freight cars, car floats, lighters, and barges. A key part of the analysis involved determining the destination or origin of each freight shipment and the time needed to deliver it to a final destination.

In addition to clockings and collecting tonnage data, the analysts determined the costs associated with each step in the freight distribution system, including those for railroad personnel, maintenance of way, equipment, train operations, capital costs, and taxes. Data collectors used a three-page form, in small print with 532 separate cost items, to capture the true costs of moving freight in the port.

The report carefully enumerated the facilities of each railroad: main lines, secondary lines, icing stations, all-rail stations, pier stations, car float terminals,

Table 5.6. Lehigh Valley Railroad facilities in 1918

Main line	Double-track freight line from Jersey City across Newark Bay and through Newark, Cranford, South Plainfield, and Bound Brook
Secondary line	Double-track belt line; National Dock Railway from Constable Hook, Bayonne to junction with NJ Junction Railroad in Jersey City, with spur to Black Tom Island
Principal yards	Break-up yard at Oak Island, Newark; waterfront yard at Jersey City; classification yard at Perth Amboy
Icing station	Jersey City
All-rail stations	Jersey City and Perth Amboy, Newark and Irvington
Pier stations	Piers 8, 34, and 66 on Hudson River; pier 44 and a pier at 124th Street; Harlem River; Wallabout Bay, Brooklyn
Float-bridge terminals	27th Street and 11th Avenue; Manhattan and 149th Street; Harlem River
Float bridges	Six at Jersey City terminal, two at Black Tom, one at 27th Street
General lighterage piers	Nine open and three covered at Jersey City terminal
Express station	Jersey City
Grain elevator	None
Coal terminals	Two car dumps and two trestles at Jersey City; trestle at Port Bayonne
Milk station	None
Ore terminal	Elizabethport
Marine equipment	Ten ferry boats, 3 steamboats, 13 tugs, 31 car floats, 16 lighters, 36 steam hoist lighters, 7 steam-derrick lighters, 44 cover barges

SOURCE: New York, New Jersey Port and Harbor Development Commission, *Joint Report with Comprehensive Plan and Recommendation* (Albany, NY: J. B. Lyon, 1920), 128–29.

float bridges, general lighterage piers, express stations, grain elevators, coal terminals, stock yards, and marine equipment. The lists for each railroad fill page after page in the report and illustrate, in great detail, the costly investment of each railroad company in an infrastructure to serve the port. The list of Lehigh Valley Railroad facilities serves as a sufficient example (table 5.6). The list for the Pennsylvania Railroad exceeded that of the Lehigh, given the size of the Pennsylvania, the largest railroad in the country. By comparison, the much smaller Philadelphia & Redding operated fewer facilities. Not only did the railroads invest significant capital; they also operated the facilities with large work forces, adding to overall costs. In addition, each railroad faced ongoing maintenance and the need for constant replacement and upgrades. Moving thousands of tons of freight exacted a toll on equipment and facilities.

Adding to costs and congestion was the fact that railroads duplicated the facilities of neighboring railroad lines, especially in the Jersey City–Hoboken–West New York corridor on the New Jersey side of the Hudson. Over the decades all of the railroads fought and schemed to bring their freight services to the shores of New York harbor and defended their franchises with tenacity. Each railroad's investment in freight yards, terminals, car floats, and tugboats represented a significant share of company's capital investment. A more efficient, integrated system to move freight would inevitably render many facilities redundant. As with all plans proposed over the decades to improve the transportation system in the port, each railroad reflexively clung to its long-standing way of doing business. Resistance to change involved more than just corporate inertia; for many of the railroads, the reorganization of the port threatened both long-term capital investments and the very existence of their freight service to the port of New York.

In addition to all of the investment in facilities and equipment, the railroads operated multiple systems to move freight between New Jersey and Manhattan or Brooklyn. The *Joint Report* described the four different ways the railroads moved freight. For miscellaneous packages and boxes, the freight remained in cars that were loaded onto car floats. Towed across the Hudson, the freight cars remained on the car floats at the railroad pier stations to be unloaded and then reloaded with returning freight. For full cars, the railroads shunted the cars off the car floats and then to freight warehouses near the waterfront. Freight destined for ships berthed in the port would be unloaded onto lighters at the freight yards and then towed across the harbor to be loaded directly onto ships at the piers on the New York side. For perishable goods, especially produce, milk, and poultry, the railroads unloaded the cars in the freight yards on the New Jersey waterfront, and loaded the goods onto horse-drawn drays that crossed the river by ferry.

A more complicated system could not be imagined, and most of the railroads employed all four ways to deliver and pick up freight. An army of clerks struggled with mountains of paperwork to keep track of all of the freight and to ensure proper billing.

To estimate total tonnages moved through the port, the commission used 1914, the most recent year with complete traffic data reported, as the base year for analysis. The twelve railroads serving the port moved 76 million tons of cargo destined within the port of New York (table 5.7). These amounts included freight transferred through the port destined for another part of the country, referred to as "interchange." For example the Baltimore & Ohio carried freight destined for New England and floated the freight from Staten Island up the East River to the New Haven Railroad facilities on the Harlem River. The freight destined for Manhattan from the New Jersey railroads totaled over 17 million tons, all of it floated across the Hudson. If the Manhattan traffic alone did not create chaos on the water, 11 million additional tons were delivered to the Brooklyn waterfront and the Queens side of the East River.

Data gathered for the Harbor Development Commission report detailed the incredible freight congestion along the Hudson River. While the New York Central took the bulk of the blame for the west side problem, most of the freight arrived over the water from New Jersey to the piers along the Hudson to be distributed in Manhattan. In just one year the New Jersey railroads moved over six million tons of freight to and from the west side of Manhattan. In addition, the railroads handled two million tons through one of six inland freight depots

Table 5.7. Railroad tonnage moved through port of New York in 1914

	Tonnage by direction of flow		
	Inbound	Outbound	Total
New Jersey railroads	50,367,000	8,006,000	58,373,000
New York railroads	2,992,000	1,750,000	4,742,000
Total	53,359,000	9,756,000	63,115,000
Interchange			12,937,000
Grand total			76,052,000

	Tonnage by destination		
	N.J. RRs	N.Y. RRs	Total
Bronx	321,000	938,000	1,259,000
Manhattan	17,884,000	2,908,000	20,792,000
Long Island	11,667,000	391,000	12,058,000
N. Jersey and Staten I.	14,764,000	47,000	14,811,000
Coastwise shipping	7,616,000	198,000	7,814,000
Export and import	6,121,000	260,000	6,381,000
Grand total	58,373,000	4,742,000	63,115,000

SOURCE: Harbor Development Commission, *Joint Report*, 140.

(table 5.8). The Baltimore & Ohio, Erie, Lehigh Valley, and Pennsylvania railroads used both pier stations and inland freight terminals. These railroads used steam engines to haul freight cars from the waterfront piers a couple of blocks inland to their freight depots, adding more congestion on West Street and along the entire waterfront. Thousands of horse-drawn drays and trucks hauled the freight the final mile through the narrow cross streets and up and down the major avenues in a never-ending stream, day and night, all year long.

To reach St. John's Park freight depot in lower Manhattan, the New York Central steam engines ran on the railroad's tracks on 10th Avenue. In 1914 the Central hauled a total of 141,290 railroad freight cars down the west side of the island to and from its 33rd Street freight yard. Further south the railroad hauled more than 60,000 freight cars between the 33rd Street rail yard and the St. John's Park freight depot in lower Manhattan. Certainly "death alley" served as an appropriate description for a freight transportation system that put 60,000 freight cars on the city's streets. When in 1846 the state of New York granted the Hudson River Railroad a charter to build a railroad connecting Poughkeepsie and New York, no one could have predicted, sixty-four years later, the volume of freight traffic on the Hudson's tracks down the west side of Manhattan, now a crucial part of the New York Central's valuable freight franchise.

The pier stations also contributed to the congestion on the Manhattan water-front. South of 60th Street on the Hudson, the railroads operated freight depots at twenty-seven piers. On the East River, the railroads used an additional seven-teen piers. Since the piers opened at only one end for both entrance and exit, incoming and outgoing freight and drays crowded West Street and South Street day and night. Adding to the street congestion, hundreds of passengers and freight wagons formed long lines waiting for ferries. Of course, all the freight arrived at the Manhattan shore by lighter or car floats. Thousands of these, towed by tugboats, filled the harbor each day, mirroring the chaos along the shore.

A crucial part of the Harbor Development Commission's analysis consisted of a detailed study of the costs incurred by the railroads moving freight in the port. In 1914 a representative sample of railroads averaged $1.60 a ton for freight handled at pier stations or inland stations on Manhattan, and $1.48 per ton for Brooklyn. For freight transported by lighters or car floats, the costs increased to $2.14 a ton. These costs could not be passed on by the railroads to shippers; the same freight rate applied to goods brought to the New Jersey side of the port as to freight delivered to a pier or inland station across the Hudson in New York or Brooklyn. Commission staff analyzed the costs to the pier and inland stations on the west side of Manhattan to get a handle on the magnitude of the west side problem. In addition, the piers on the Hudson were among the most crowded in the entire port. The commission estimated costs on a per car basis, including

Table 5.8. Tonnage and cars handled south of 60th Street, 1914

	Tonnage			No. of freight cars		
	Inbound	Outbound	Total	Inbound	Outbound	Total
Pier stations						
South of 30th St.	2,525,885	1,511,385	4,037,270	328,760	328,760	657,520
North of 30th St.	193,164	143,260	336,424	21,933	21,933	43,866
Total	2,719,049	1,654,645	4,373,694	350,693	350,693	701,386
Inland station (railroad)[a]						
St. John's Park (NYC)	213,484	107,286	320,770	30,558	30,558	61,116
33rd St. (NYC)	753,788	205,789	959,577	70,645	70,645	141,290
25th St. (B&O)	74,465	32,393	106,858	7,460	7,460	14,920
26th St. (Lehigh)	23,760	14,160	37,920	3,000	3,000	6,000
28th St. (Erie)	99,486	44,369	143,855	8,315	8,315	16,630
37th St. (Penn.)	108,172	64,711	172,883	13,710	13,710	27,420
Total	1,273,155	468,708	1,741,863	133,688	133,688	267,376
Grand total	3,992,204	2,123,353	6,115,557	484,381	484,381	968,762

SOURCE: Harbor Development Commission, *Joint Report*, 152.
[a] NYC = New York Central; B&O = Baltimore & Ohio.

Table 5.9. Operations costs to move freight to Hudson River piers (from breakup yard to station floor), 1914

Operation	Direct cost	Fixed cost	Total
Classification at breakup yard	$0.56	$0.19	$0.76
Breaking bulk at transfer platform	2.72	0.04	2.76
Line haul between yard and terminals	0.41	0.28	0.69
Switching and bringing to waterfront	1.01	0.87	1.89
Car float operations	1.32	0.28	1.61
Handling freight at terminal	3.89	2.10	5.90
Total	$9.84	$3.78	$13.62

SOURCE: Harbor Development Commission, *Joint Report*, 164.

both direct operating costs and fixed charges derived from the large investment in facilities and equipment (table 5.9).

Steamships

A second part of the commission's study focused on the shipping that crowded the port's water and piers. Ships carried 45 million tons of freight that entered or left the port of New York in 1914. New York's harbor had for centuries drawn the world's shipping, and its piers remained the center of maritime commerce in the United States. Transferal of freight between ship and railroad, based on an elaborate and costly system developed in piecemeal fashion, created a whole other tier of complications in the port.

To measure the time and effort involved in both loading and unloading, commission researchers took a sample of arriving and departing ships and sent teams to conduct clockings, carefully recording all of the operations necessary, including the number of men employed and their wages. Unloading an arriving ship involved three steps: breaking out cargo in the hold, hoisting the cargo out of the hold and onto the pier, and stowing freight on the pier or loading it directly onto a lighter. Loading reversed the process. Stevedores and longshoremen worked the piers and ships in the harbor, performing back-breaking labor for long hours, a tough, often brutal occupation.

Loading and unloading operations varied dramatically by the type of ship (oceangoing vs. coastal, for example), and especially with the type of cargo. To make sure the clockings represented shipping to all parts of the world, the researchers divided shipping into fourteen zones:

1. United Kingdom and Europe
2. the Mediterranean
3. Africa

4. India
5. the Far East and the east coast of Asia
6. the East Indies and Australia
7. the north and east coasts of South America
8. the west coast of South America
9. the West Indies and the east coast of Central America
10. miscellaneous Atlantic
11. miscellaneous Pacific
12. domestic coastwise shipping
13. service to Long Island Sound and other local salt-water lines
14. lines on the Hudson River[54]

The list illustrates the incredible web of shipping linking the port of New York to the entire world. No other port in the United States or in the world offered the shipping connections of New York. Because it provided service all over the world, numerous companies competed for freight business. Twenty-two steamship companies offered freight service to "Zone 1" (Great Britain and Europe), including the famous Cunard Line, Holland-American, the North German Lloyd, and the Swedish-American Line. Four lines served Africa, sixteen South America (including W. R. Grace & Co. and United Fruit), and six shipped freight to Asia, including Australia.

In the period from 1907 to just after the end of World War I in 1919, the number of ships entering and leaving the port grew. Commission researchers gathered information from the Customs Service on the number of ships passing through the port in "foreign-direct service" (table 5.10). Foreign-direct meant a ship arriving or departing New York straight from or to a foreign destination with no other port of call in the United States.

For this time period an average of over 4,000 ships entered the port of New York each year in foreign trade; eleven ships a day arrived, and on average nine departed. The study reported that an additional 3,796 ships cleared the harbor for the coastal trade in the United States in 1919; this number did not include Hudson River and Long Island Sound vessels. For each of these years the port of New York handled the greatest concentration of shipping in the world. All the major east coast railroads came to the shores of the harbor, constructed freight yards and piers, and operated lighters and tugboats, all at great expense, drawn by the sheer magnitude of shipping in the port.

The costs for loading and unloading varied widely from ship to ship and by the type of cargo carried. To complicate costs, the efficiency of moving cargo depended on how up-to-date a particular pier happened to be. The city leased the piers to private shipping companies. Private companies also

Table 5.10. Foreign-direct shipping from the port of New York

Year	No. of ships	
	Entered	Cleared
1907	4,315	3,863
1908	4,011	3,274
1909	4,154	3,134
1910	4,149	3,204
1911	4,081	3,416
1912	4,095	3,347
1913	4,448	3,706
1914	4,200	3,478
1915	4,783	4,295
1916	5,192	4,837
1917	4,630	4,165
1918	4,040	3,326
1919	5,016	4,316

SOURCE: Harbor Development Commission, *Joint Report*, 175.

owned their own piers, and the privately owned piers tended to be the more modern ones. The commission reported costs per ton for loading and unloading ranged between $0.97 and $3.17.

Shipping efficiency also depended upon the number of days a ship spent in the port, referred to as "lay days." Transatlantic ships averaged five or six days in port; Long Island Sound ships, two or three days. A ship in port earned nothing for the owners; consequently, ship owners often paid significant overtime charges to get their vessels loaded or unloaded as quickly as possible.

Increasing the efficiency of loading and unloading cargo would speed shipping in and out of the port but would not solve the problem of moving freight between the ships, piers, and railroads. To deal with the latter problem, the Harbor Development Commission recommended an elaborate plan, an "automatic-electric system" to move freight between New York and New Jersey. While the commission referred to the plan as "innovative," many of its essential features mirrored Wilgus's plan for the small-car freight subway proposed to the Public Service Commission twelve years earlier.

The Plan Recommended—the Automatic-Electric System

The Harbor Development Commission recognized that the key to streamlining the freight distribution problem in the port of New York was a vastly improved system for moving freight between New Jersey and Manhattan. The commission proposed to use a railroad system, with two tunnels under the Hudson River, a new rail yard in the Hackensack Meadows, and a deep underground freight subway in Manhattan. The plan called for automatic eight-car freight trains to run continuously, in one direction, from the rail yard in New Jersey through the north tunnel under the Hudson and then deep under the streets of Manhattan, stopping under twelve freight terminals from 60th Street to the Battery. At each terminal, freight to be locally delivered would be removed and outgoing freight loaded. The electric trains would then continue through the southern tunnel back to the train yard in the Hackensack Meadows. In the train yard in New Jersey, the process would start all over again. The authors of the plan described it as "the conversion of the railroad into what is closely akin to a conveyor system."

The plan assumed that all of the New Jersey–based railroads would use the new joint freight yard to transfer freight from railroad cars to the automatic electric cars. To enable all of the railroads to connect to the joint yard, the plan included the construction in New Jersey of an inner belt railroad from a junction with the Lehigh Valley Railroad in the south to the West Shore Railroad in the north.

The automatic-electric system would be centrally run without a separate electric engine, eliminating the need for an engineer to drive the train, a significant cost saving. In fact, the plan projected a number of cost savings: "reduced size of terminals, the reduction in Manhattan trackage, the reduced costs per foot of deep tunnels, and the low operating cost due to automatic operation."[55] Esti-

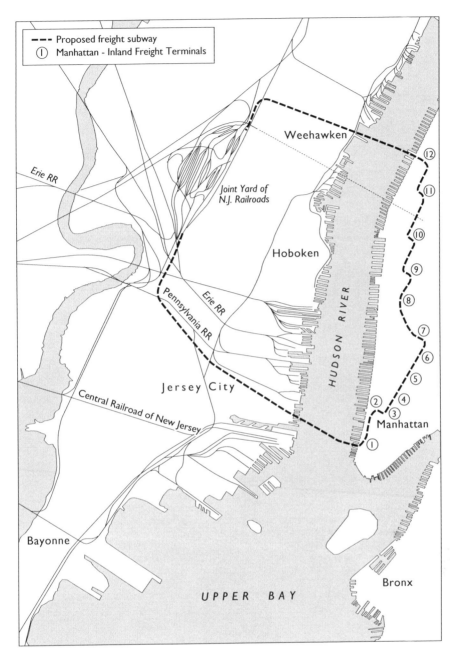

Legend:
- --- Proposed freight subway
- ① Manhattan - Inland Freight Terminals

Erie RR

Joint Yard of N.J. Railroads

Weehawken

Hoboken

Pennsylvania RR

Erie RR

HUDSON RIVER

Jersey City

Central Railroad of New Jersey

Manhattan

⑫ ⑪ ⑩ ⑨ ⑧ ⑦ ⑥ ⑤ ④ ③ ② ①

Bayonne

UPPER BAY

Bronx

Harbor Development Commission's proposed freight subway from rail yard in New Jersey to inland freight terminals in lower Manhattan, 1919.

mates of the cost, including equipment, totaled $201,290,390 with interest on the borrowing at 5 percent (table 5.11).

The Harbor Development Commission projected the volume of freight shipped on the new system to total at least 10 million tons a year. Estimates were that the automatic-electric system would reduce freight costs by $0.43 a ton, creating a total savings of $4,318,090, or 3.57 percent on the total investment of $201 million needed to construct the new system (table 5.12).

Significant savings would be gained from eliminating "car detention" fees. Under the existing system, if one railroad held the freight car of another railroad while the freight was being delivered and unloaded, the first railroad paid the second $0.75 a day for the "detention." With the crowded, convoluted system in operation in the port of New York, moving a freight car from New Jersey to New York and back averaged three and half days. The commission's plan included saving 1.8 million detention days a year, producing savings in excess of $1.3 million.

Additional savings would be realized by Manhattan-based shipping companies that hauled freight through the streets of the city. Each day the crush at the pier stations and inefficient inland freight depots created delays for the shipping companies, increasing costs. With ten terminals in lower Manhattan the automatic-electric system would decrease delivery and pick-up distances in the city, easing traffic congestion on the crowded city streets of lower Manhattan.

The Harbor Development Commission report summarized all of the projected savings and included a list of additional savings to which it could not attach dollar figures (table 5.12). In this, the commission followed the Wilgus planning model, including both direct savings and the often substantial but difficult-to-estimate indirect savings, such as reduced food spoilage and reduced street maintenance in New York.

The commission report went into further detail discussing at great length additional benefits to be derived from the plan for the automatic-electric freight system. Sections of the report summarized the favorable impact of the plan on

Table 5.11. Estimated costs for the automatic-electric system to move freight between New York and New Jersey

Item	Cost
Construction	
New Jersey joint rail yard	$13,369,605
Belt line in New Jersey	3,762,040
Lackawanna & Erie interchange yard	411,400
Southern Hudson River tunnel	21,057,625
Northern Hudson River tunnel	10,316,275
Tunnels in Manhattan	59,618,275
Twelve Manhattan freight terminals	56,308,760
Electric installation	13,610,000
Equipment	9,452,200
Total	$187,906,330
Construction interest, 5%	13,384,060
Grand total	$201,290,390

SOURCE: Harbor Development Commission, *Joint Report*, 251.

Table 5.12. Harbor Development Commission's estimate of savings from the automatic-electric system

Summary of projected savings

Savings over present railroad operations	$4,318,000
Savings in truck delays	1,648,000
Savings in trucking distances	1,154,000
Total	$7,120,000
Percentage above 5% interest on $201,290,000	3.54%

Additional savings not practical to estimate

Increased average loading of railroad cars

Increased average loading of trucks

Savings to public: quicker terminal

Reduction in trucking expenses

Reduction of trucking of milk, produce

Savings through less food spoilage

Decreased maintenance of city streets

Freedom from interruption by fog, ice

Increased real estate values

Increased commercial capacity in Manhattan

SOURCE: Harbor Development Commission, *Joint Report*, 19.

trucking, warehousing, food distribution, improvements on the Manhattan and New Jersey waterfronts, hauling grain, and even supplying ice to the city. In a concluding statement the commission urged the adoption of the system as "a solution of the West Side and Manhattan freight distribution problem." Noting that "the automatic-electric system is capable of expansion throughout the Port," the commission emphasized that solving the west side freight problem represented the key: "The solution of the entire Port problem depends upon a correct solution of the Manhattan problem."[56]

Both Wilgus and the Harbor Development Commission focused their planning efforts exclusively on a railroad solution to the port's freight distribution problems. Nowhere in the voluminous *Joint Report* does there appear any serious investigation of using highways and trucks to move freight. Whether the solution was the small-car freight subway Wilgus proposed in 1907 or the automatic-electric system the commission presented in 1919, the efficient, cost-effective way to move freight was by railroad, as it remains to the present day.

The Birth of the Port Authority

The last section of the Harbor Development Commission report of 1919 called for the creation of a permanent, bi-state commission to carry through the recommendations of its study. The commission went one step further and included a draft of a "Supplementary and Amendatory Treaty between New York and New Jersey" to establish a port authority. The proposed treaty included seventeen articles defining the functions of the new authority and two proposed statutes to be passed by the New York and the New Jersey legislatures. Directly from these proposed legal documents arose the Port Compact of 1921 establishing the Port of New York Authority.[57] Just two years later, New York and New Jersey passed the necessary legislation, and on April 13, 1921, seven commissions signed the compact officially establishing the Port Authority. Created specifically to solve the railroad freight transportation problems in the port of New York, the new Port Authority soon dramatically expanded its mission.

Over the next decades the Port Authority became a behemoth, building bridges, tunnels, highways, airports, and office buildings, none of which were part of the plans proposed by the Harbor Development Commission in 1919. In a 1961 article Sidney Goldstein, then the general counsel of the Port Authority, boasted that the authority currently operated "twenty-two terminal and transportation facilities in the metropolitan region, . . . an area composed of over 13,000,000 people residing in more than 360 separate and distinct communities."[58] At that time the Port Authority investment totaled over $1 billion. The list of facilities included four bridges, two tunnels, four airports, six marine terminals, three inland freight terminals, and a bus terminal. The 1961 list does not include a single railroad facility—the original purpose for establishing the authority in the first place.

In the 1960s the Port Authority planned to open a third two-lane tunnel addition to the Lincoln Tunnel at a cost of $95 million, add a lower deck to the George Washington Bridge, and build a bus terminal in midtown Manhattan, to cost $180 million. Future plans called for "a new bridge across the Narrows of New York harbor, linking Staten Island to Brooklyn . . . to improve a system of vehicular bridges and highways."[59] By the early 1960s, the authority's annual reports illustrate the triumph of the highway, truck, and automobile in the port of New York.

Despite the fact that no credit accrued to Wilgus for the original idea of the Port Authority, he supported its establishment in 1921. As president of the New York section of the American Society of Civil Engineers, he issued public statements urging the New York State Legislature to pass the bill creating the authority.[60] The influential Merchant's Association of New York also lent its voice in support.

But not all interests in the port favored the establishment of an authority that was, for all intents and purposes, outside the exiting political structure. The mayor of Newark vigorously opposed the authority as a threat to local control and claimed it would "build up for itself the greatest concentration of power and property that has ever existed anywhere in the world."[61] In the view of New York's Mayor Hylan, the power delegated to the Port Authority came at a loss to the city of New York. In the future, major decisions about transportation would be made by appointed, not elected, commissioners insulated from the public and locally elected officials.

Wilgus objected to the final form of the Port Authority, especially the fact that the commissioners appointed by the governors of New York and New Jersey did not possess the necessary transportation expertise. In a number of spirited letters to Julius Cohen, general counsel of the Port Authority, written in August 1925, he voiced his concerns. First Wilgus sent Cohen a thick docket of material from his files for the time period from July 1909 through February 1911 to document his claim of having originated the idea for a two-state authority.[62] Wilgus next identified the key problem with the Port Authority: "I have been convinced that such a powerful body . . . should be non-political and truly representative of the interest whose property should be incorporated in the modernized port. . . . These interests comprise the rail and water carriers, the rate payers (shippers), the autonomous communities embraced within the port."[63] He added that the authority's chief executive needed to be "of the highest executive ability and widest experience in transportation." Perhaps Wilgus had himself in mind.

Cohen and Wilgus continued to exchange long letters over the mission and organization of the authority. At one point Cohen sent Wilgus a twenty-five page letter in response.[64] Wilgus argued for the Port Authority to be modeled after the successful port authorities recently established in London and Liverpool, which included wide representation from shipping and railroad interests in the two ports. Cohen did not concede the point and defended the current governing structure of the authority, in which the six commissioners were appointed by the two governors. He forcefully rejected "the delegation of sovereign power to railroads and shippers and steamship owners doing business at the Port of New York. In our country we do not do things that way."[65] Wilgus and Cohen sharply disagreed on the appropriate governing structure for the Port Authority.

Cohen did offer Wilgus a back-handed compliment after reviewing the portfolio of documents supporting his authorship of the idea for a portwide authority: "I find the papers which you lent me showing your connection with the ideas of a general scheme for interstate port development, fascinatingly interesting. It seems to me exceedingly strange that, during the formative period when the legislation of the two states was passed . . . and afterwards, while the commission was at work, your opinions never came to our attention as far as I can recall. Perhaps it was because you were on the other side."[66] The "other side" referred to Wilgus's service with the AEF in France during the crucial period leading up to the creation of the Port Authority. For a second time in his career Wilgus adamantly believed that proper credit for an extraordinary idea had passed from him to others.

From Railroads to Other Modes of Transport

When the Port Authority finalized its first comprehensive plan and both state legislatures endorsed the plan, the authority expected cooperation from the railroads. But just as Wilgus had encountered resistance from the railroads to his various plans to solve freight problems, the Port Authority found it impossible to get the railroads to participate in the plan. Jameson Doig describes the ensuing

Table 5.13. Facilities built or acquired by the Port Authority, 1927–1950

Facility	Year built or acquired	Acquisition information
Holland Tunnel	1927	Taken over by Port Authority in 1930
Goethals Bridge	1928	
Outerbridge Crossing	1928	
George Washington Bridge	1931	
Bayonne Bridge	1931	
Inland Terminal No. 1	1932	
Lincoln Tunnel, South Tube	1937	
Grain Terminal and Columbia Street Pier, Brooklyn	1944	
Lincoln Tunnel, North Tube	1945	
Port Newark and Newark Airport	1947	Acquired under lease from the city of Newark
LaGuardia and Idlewild Airport	1947	Acquired under lease from the city of New York
Teterborough Airport	1948	
New York Truck Terminal	1949	
Newark Truck Terminal	1950	

SOURCE: John Griffin, *The Port of New York* (New York: City College Press, 1959), 76.

conflict as "Ten Years of Travail" and points out the railroads rejected plans for cooperation and joint improvements for decades.[67] The ever-present Ira Place, general counsel of the New York Central, argued that "as long as we are engaged in private business . . . there is only one factor which gives the public a return and that is self-interest."[68] Adam Smith would have smiled. Certainly, no one ever accused the railroads of not acting in their own narrow self-interest. Rational planning of the type practiced by both Wilgus and the Port Authority failed whenever the railroads perceived their own vested interests to be at risk.

The resistance of the railroads to the authority's first comprehensive plan pushed the agency to devote its energy to other, competing modes of transportation. The Port Authority enthusiastically embraced the highway revolution. Trucks soon captured the major share of freight transportation in the port of New York, making intensive use of the authority's new bridges and tunnels and sounding the death knell for the private railroads serving the port, all of whom marched inexorably toward bankruptcy in the next decades.

A 1959 study listed the facilities acquired or constructed by the Port Authority between 1927 and 1950 (table 5.13). Conspicuous by their absence are major improvements in rail facilities. The success of the Port Authority in completing a stunning number of complex projects rested on its ability to issue bonds to finance construction based on the promise that all user revenue would be devoted exclusively to servicing the bonds. With tolls and fees, increasingly from automobiles, trucks, and eventually the airlines, more than sufficient to cover the cost of all borrowing and maintenance, the ever-increasing surplus could be employed for the next round of projects. Eventually, the authority built the largest office buildings in the world at the time, the twin towers of the World Trade Center.

From the end of World War II, the Port Authority prospered. Revenue from facilities more than covered the cost of borrowing and left a significant surplus the authority could tap into to build and expand its reach. For example, in 1945 the authority's gross operating revenue of $19.3 million, after covering all operating costs, left a net revenue of $13.3 million, which covered the funded debt of $183.1 million. Five years later, funded debt had increased to $248.4 million, but revenue more than doubled to $42.2 million, leaving a net revenue of $29.2 million. By 1957 gross revenue doubled again to $84.7 million, easily supporting funded debt totaling $420.6 million.[69]

The Port Authority created a virtual self-financing cornucopia machine with tolls from bridges and highways and later from landing fees at the airports, providing more than enough revenue to cover all costs and leaving a huge surplus to be invested in new projects. All of this fiscal power wielded by an independent

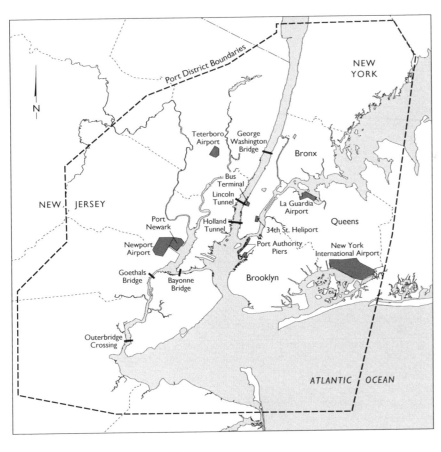

The major facilities operated by the Port Authority by 1950: four airports, four bridges, two Hudson River tunnels, a bus terminal and a heliport in Manhattan, the port of Newark, and Brooklyn piers.

bi-state authority with six non-elected commissioners and a smart, aggressive professional staff remained, to a large degree, insulated from political pressure from either the mayor of New York or any other elected officials within the port district.

Over the years, the authority battled tooth and nail to resist efforts to use its ever-increasing revenue to support the region's crumbling commuter railroads and mass transit operations. Eventually, intense political pressure forced the Port Authority to take over the operations of the bankrupt New Jersey & Hudson Railroad Company and operate the passenger subway tubes under the Hudson River. Reconstituted as the New York–New Jersey PATH lines, the service provides a crucial commuter link between New Jersey and Manhattan. Periodically, the Port Authority complains about subsidizing the money-losing PATH system and being forced to divert revenue from its immensely profitable bridges, tunnels, and airports.

When Wilgus and others began arguing for a port of New York authority, they could not have imagined the impact of the highway and airport age. Writing in 1961, the general counsel of the Port Authority described with great enthusiasm the authority's two new "heliports" that had been opened in Manhattan. On the other hand, the Port Authority delayed constructing mass transit rail links to the airports, as every other major airport in the world had done.

When the authority opened a new passenger terminal at Idlewild (now Kennedy Airport) in 1960, they christened it "Terminal City." Forty-seven years earlier, the New York Central Railroad had proudly referred to its new Grand Central project as New York's "Terminal City." In 1960, the once-magnificent Grand Central Terminal and the soon-to-be-destroyed Pennsylvania Station no longer served as gateways to the greatest city in the world. Visitors now arrived by airplane, and the authority spent a great deal of money extending the length of the runways to accommodate the first jet airplanes, providing service to the entire world, as the port's shipping companies once did.

Throughout the 1940s and 1950s the Port Authority undertook initiatives to improve the shipping facilities in the port, especially with the development of Port Newark in New Jersey. The port's shipping increasingly left the Manhattan

Table 5.14. Total activity at New York and rival domestic ports, 1935–1955 (millions of tons of cargo)

Year	New York	Boston	Philadelphia	Baltimore	Norfolk	New Orleans
1935	97.1	16.4	20.8	17.3	14.2	12.9
1940	129.8	19.0	29.8	26.9	21.6	19.8
1945	108.6	15.0	37.1	32.2	16.0	24.2
1950	144.9	19.4	52.6	35.6	18.2	35.1
1955	148.1	19.1	67.6	45.8	35.9	47.1

SOURCE: Griffin, *Port of New York*, 46.

and Brooklyn piers for the much more modern facilities at Port Newark. Freight continued to shift from the railroads to trucks using the new bridges, tunnels, highways, and truck freight terminals. Even with the switch from rail to road, the port of New York continued to flourish. Although New York's rival ports increased their shipping, as late as 1955 New York still handled the bulk of the country's imports and exports, with the freight increasingly hauled by truck (table 5.14).

Despite New York's lead, other ports, particularly Philadelphia and Baltimore, posed increasing challenges. The cost of freight handling in the port of New York remained high. Obsolete piers in Manhattan and Brooklyn could not be modernized, especially for container operations, given the lack of land for expansion. One study of the

Table 5.15. Comparison of freight handling costs in New York and New Orleans for a C1 type vessel loading 2,600 tons of general cargo, 1954

Operating charges	New York	New Orleans
Warfage and dockage	$1,924	$384
Stevedoring	7,278	5,278
Checking cargo	1,222	520
Watchmen	468	312
Labor receiving	912	416
Tug hire lighters	234	
Supervision	260	104
Timekeepers	52	26
Miscellaneous	208	234
Total	$12,558	$7,274
Cost per ton	$4.83	$2.79

SOURCE: Griffin, *The Port of New York*, 124.

costs of handling provided a detailed analysis comparing New York with New Orleans (table 5.15). The port of New Orleans operated with a cost advantage of almost two dollars per ton. In addition, these costs represented only moving freight from a ship onto the adjoining pier. In New Orleans railroad tracks lined many of the piers, enabling cargo to be quickly loaded into freight cars. On the New York side of the harbor, freight still had to be handled multiple times, floated across the Hudson River, unloaded and then reloaded again before a single ton left the port.

New York's cost differential could not be sustained in the long run. The only solution was to move ships and cargo away from the piers in Manhattan and Brooklyn, where the port of New York had flourished for centuries. A revolution in cargo handling began that forever changed shipping in New York and around the world—containerization. Shipping freight in already loaded containers on specially built ships and unloading the containers without any transfer of freight dramatically reduced loading and unloading time and costs. The number of workers on the aging docks in Manhattan and Brooklyn vastly exceeded the skeleton crews needed to operate the huge container ports.

Bowing to the inevitable, the Port Authority turned its attention to the New Jersey side of the harbor, where land available on the shore offered the space needed to build new shipping facilities. In 1946 the Port Authority entered negotiations with the city of Newark, New Jersey, to have the authority take over Port Newark, situated on Newark Bay. In May 1947 the city of Newark agreed to lease to the Port Authority, and the authority immediately began a program of modernization. By 1957 the authority's investment in Port Newark had reached $27 million. The port provided shipping with modern, up-to-date container facilities and significantly lower operating costs. Railroads had easy access to the piers, and as importantly, modern highway connections allowed access to the piers by trucks. Overall, the improvements in Port Newark helped the port to remain competitive but spelled the end for freight and shipping back across the harbor in New York.

Not completely satisfied with improvements to Port Newark, the authority made additional investments. In 1958 the Port Authority, with great fanfare, announced plans for a "huge new $150,000,000 marine terminal on Newark Bay in Elizabeth, New Jersey."[70] Just to the south of Port Newark, this new facility offered the most up-to-date facilities and eventually, along with Port Newark, switched to container operations.

The Port Authority never completely abandoned shipping operations in New York. In the 1950s it made a last effort to improve freight handing along the Brooklyn waterfront, investing over $85 million to improve the docks just to the south of the Brooklyn Bridge. In Manhattan the city Department of Marine and Aviation, successor to the Department of Docks, labored to make improvements to the 134 city-owned piers. Not to be left out, the Port Authority proposed in 1949 to lease all of the city-owned piers for a period of fifty years, undertake all needed improvements, and then sublease the piers to the shipping companies and the railroads. Always quick to guard local control, the city of New York rejected this offer.

In the 1960s and 1970s, shipping abandoned the storied piers lining the shores of Manhattan and Brooklyn. Occasionally a cargo ship could be glimpsed tied up to a pier in Brooklyn or a cruise ship, the new *Queen Elizabeth*, might be seen arriving with enormous fanfare to berth at a Hudson River pier. Today almost all of the shipping activity in the port of New York takes place on the New Jersey side of the harbor. Once, the railroads and New Jersey politicians and businessmen complained vociferously about the competitive advantages enjoyed by the city of New York. No longer; the long, slow death of freight and shipping to the "island at the center of the world" ended without any great fanfare. The lifeblood of the port now beats at the vast container port in Newark Bay. No action on the part of the city of New York or the Port Authority or the railroads could have forestalled the inevitable end of Manhattan's position as the world's shipping center, a role the island enjoyed for over two centuries.

Even if Wilgus's small-car freight subway or the Inter-Terminal Belt Line had been built, the eventual decline of Manhattan as a shipping center would, in all probability, have followed. New transportation technologies demanded more space and greater efficiency than the cramped streets and piers in Manhattan provide.

Established to solve the port's freight problems, the Port Authority first turned away from the railroads as the means to improved transportation when it could not secure any meaningful cooperation from the railroad companies. It then moved the port's shipping from New York City to Newark Bay. With a flood of revenue pouring in each year, the Port Authority secured the financing and had the power to effect changes to the port of New York on a monumental scale, and it seldom hesitated to use its power. Wilgus persisted in claiming credit for originating the idea of a bi-state authority to plan and improve the port, and he continued to believe the railroads to be the backbone of the region's transportation system. He never imagined how the authority that he first suggested in 1907 as a means to solve the railroad freight problems would shape the future of the port of New York and the entire metropolitan region.

6

MAKING ROOM FOR THE AUTOMOBILE

The Holland Tunnel

TUNNELING UNDER RIVERS began in England in 1842, when Marc Isambard Brunel, after over nine years of effort, completed a tunnel under the Thames River in London. Brunel patented his "shield" method for constructing a tunnel under a riverbed through mud and silt. The shield consisted of a covered space for workers to excavate, a system to remove materials, and hydraulic jacks to push the shield forward. English engineers added compressed air and airlocks to keep water out and the use of cast-iron rings to line the wall of the tunnel as the shield moved forward.[1]

The first use of the shield method in the United States was in 1869, when A. E. Beach, editor of *Scientific American*, supervised the construction of Manhattan's first subway, which covered just a short distance under Broadway. The cars in this subway tunnel, which was a mere eight feet in diameter and lined with brick, were pushed forward by compressed air.[2] In the 1870s, American engineers began to use the shield tunnel method for railroad tunnels under rivers through the silt and mud of the riverbed. Borrowing from the English methods, "subaqueous" tunneling employed the shield technique with compressed air pumped in to keep the water out. Workers entered through an airlock and excavated material at the front of the shield. As the excavation continued, workers bolted new rings on to extend the tunnel walls. Hydraulic rams forced the shield forward as the tunnel advanced under the riverbed. Engineers developed variations to deal with different kinds of mud and silt encountered in riverbeds. One method, for especially porous silt, used a series of small doors in the shield at the forward end. The doors could be opened, allowing silt and mud to enter the completed section of the tunnel and be removed. With enough pressure, the hydraulic rams at the rear created the necessary force to push the shield forward.

Bridge builders also used compressed air to keep water out as they dug foundations for bridges over rivers. In New York, the Roeblings employed caissons, airlocks, and compressed air to construct the foundations for the two towers of their world-famous Brooklyn Bridge, which opened on May 24, 1883. The two towers stood in the river, and the engineers needed to have the them rest on bedrock deep under the bottom of the East River. In this case, the caissons descended vertically through the mud and silt to reach bedrock. For a tunnel under a river, the shields moved horizontally under the riverbed through the same mud and silt.

The First Hudson River Passenger Tunnels

In 1879, DeWitt C. "Colonel" Haskins, a Civil War veteran and railroad engineer who had gained experience working for the Union Pacific Railroad, announced plans to construct the first subaqueous shield tunnel in the port of New York. He planned to construct a railroad tunnel under the Hudson River linking Jersey City to lower Manhattan at Morton Street. During the "age of energy," engineers

Diagram of New York and New Jersey Trolley Tunnel

Working on the Tunnel in the Compressed-air Shaft

Illustrations of the first tunnel under the Hudson River, providing rail passenger service between Manhattan and Hoboken, New Jersey, 1902.

Tunnel boring shield excavating a tunnel under the Hudson River for the Hudson & Manhattan Railroad, 1908.

like Haskins never hesitated to take on daunting challenges. After New York and New Jersey passed the necessary enabling legislation, Haskins formed the Hudson Tunnel Railroad Company, raised capital, and began construction. After many delays, legal battles with the railroads, and spending millions of dollars, Haskins stopped work on the tunnel in 1892. With over two thousand feet of tunnel completed, Haskins had clearly demonstrated the viability of subaqueous tunneling under the Hudson River.

In 1902, the brilliant engineer and entrepreneur William McAdoo took over the Hudson River tunnel project. He reorganized the business as the Hudson & Manhattan Railroad and completed two tunnels in just six years. Rapid transit service between Jersey City and Cortland Street in Manhattan opened in 1908 and proved to be an immediate success. Decades later, the Port Authority took over the Hudson & Manhattan Railroad and continues to operate the tunnels as part of the PATH system.

In effect, the Hudson & Manhattan had constructed a "subway" under the Hudson River, connecting Jersey City to Manhattan with the electric-powered cars, modeled after the equipment used on the IRT. Tens of thousands of daily commuters who lived in New Jersey but worked in Manhattan could now ride the subway cars of the Hudson & Manhattan.

With the technology of subaqueous tunneling established, the Pennsylvania Railroad decided to use the technology to achieve a long-sought-after goal—direct rail access to Manhattan Island. After endless political intrigue, the railroad began construction in 1905 of two tunnels under the Hudson River from Jersey City whose tracks led to the new Pennsylvania Station at 33rd Street.[3] Opened for service in 1909, the tunnels allowed the passenger trains of the Pennsylvania system to travel from afar directly onto Manhattan Island and allowed the Pennsylvania to challenge the New York Central's monopoly on direct-passenger service to New York. Now two railroads offered direct passenger service to Manhattan.

For a period of time, the subaqueous tunnels under the Hudson were the longest under-river tunnels in the world. As is often the case, transportation construction projects in the United States required a much larger scale. Brunel's first tunnel under the Thames in London, from Wapping to Rotherhithe, totaled 1,506 feet in length. A second tunnel in London, under the Thames between Tower Bridge and London Bridge, was 1,320 feet.[4] By comparison, the Hudson & Manhattan tunnels totaled 3,507 feet. The Pennsylvania Railroad's tunnel was over 6,000 feet in length, far longer because it started on the west side of Bergen Hill in New Jersey and not on the New Jersey waterfront.

B. H. Hewitt, the engineer who directed the completion of the Hudson & Manhattan tunnel, described the crucial role the engineer played: "In tunneling with a shield, more so than in any other branch of engineering, it is the engineer's function to direct the work . . . even down to the smallest details. . . . He will be charged with a never ceasing load of cares that cannot be left to the contrac-

Bergen Hill, New Jersey, entrance to the Pennsylvania Railroad tunnel under the Hudson River, 1912.

The Pennsylvania Tunnels, New York.

tor. . . . Every move of the shield is an experiment which must be controlled and directed by the engineer."[5]

While both the Hudson & Manhattan Railroad and the Pennsylvania Railroad tunnels provided rail links connecting Manhattan to New Jersey, both operated only passenger trains through the tunnels. Moving freight within the port still required fleets of lighters, barges, and tugboats. In spite of the expense and efforts undertaken to complete these first tunnels, the delivery of freight still involved crossing the Hudson River by water.

Continued Demand for Hudson River Crossings

Even as work progressed on the first passenger tunnels under the Hudson River, the clamor for additional crossings over or under the river continued. Pressure arose from many quarters. Politicians and commuters on both sides of the river wanted quicker and more reliable transportation than the ferries provided. Even with fifteen ferry lines and multiple ferries operating around the clock, inevitably delays occurred as passengers and vehicles crowded the ferry terminals along the shoreline waiting to board. Numerous companies operated short-haul freighting between the rail yards on the New Jersey shore and lower Manhattan, using the ferries for their horse-drawn drays or, later, the first motor trucks. This system only added to the crush of traffic at the ferry piers. Most ferries used the upper deck for passengers and the lower deck for vehicles. During the hot summer days, passengers often complained of the smell of horse manure.

Table 6.1. The two major ferry groups between Manhattan and New Jersey, 1919

	Departure points	
	New Jersey	Manhattan
Five-ferry group		
Pennsylvania Railroad	Exchange Place, Jersey City	Desbrosses St.
Pennsylvania Railroad	Exchange Place, Jersey City	Cortlandt St.
Erie Railroad	Pavonia Ave., Jersey City	Chambers St.
Delaware, Lehigh & Western	Hoboken	Barclay St.
Delaware, Lehigh & Western	Hoboken	Christopher St.
Six-ferry group		
Central Railroad of NJ	Communipaw Ave., Jersey City	Liberty St.
Central Railroad of NJ	Communipaw Ave., Jersey City	W. 23rd St.
Erie Railroad	Pavonia Ave., Jersey City	W. 23rd St.
Delaware, Lehigh & Western	Hoboken	W. 23rd St.
Delaware, Lehigh & Western	14th St., Hoboken	W. 23rd St.
West Shore Railroad	Weehawken	Cortlandt St.

SOURCE: State of New York, *Report of the New York State Bridge and Tunnel Commission*, Legislative Document no. 24 (Albany, NY: J. B. Lyon, 1921), 17.

Table 6.2. Average weekday traffic for five-ferry group traffic, 1914–1920

	No. of vehicles			% increase	Total annual passengers
	Horse-drawn	Motor-driven	All		
1914	7,655 (86.9)[a]	1,159 (13.1)	8,814	—	2,839,055
1915	7,137 (82.7)	1,497 (17.3)	8,634	2.0	2,811,249
1916	7,218 (78.0)	2,038 (22.0)	9,256	7.1	3,042,756
1917	6,785 (68.7)	3,083 (31.3)	9,878	6.9	3,279,019
1918	6,270 (59.9)	4,193 (40.1)	10,463	5.8	3,496,620
1919	5,693 (50.9)	5,404 (49.1)	11,177	6.8	3,731,032
1920	4,994 (43.4)	6,509 (56.6)	11,503	2.9	3,853,793

SOURCE: State of New York, *Report of the New York State Bridge and Tunnel Commission*, 18.
[a]Numbers in parentheses, percentages.

To respond to the demands for additional tunnels or bridges, the state of New York in 1906 created the New York Interstate Bridge Commission to study the feasibility of building a bridge over the Hudson or more tunnels. New Jersey followed with a similar legislation, and the two states formed a joint commission. The New Jersey legislation clearly set forth the mandate: "The necessities of interstate commerce and convenience and safety of people residing in the neighborhood of the Hudson demand a more modern, rapid and economical system of transit to the city of New York than by the slow, antiquated system of ferry boats."[6] From the joint commission's perspective, the immediate need for improved transit did not include the entire port of New York. The commission focused on the local traffic between lower Manhattan and the New Jersey shore directly across the Hudson in Jersey City and Hoboken.

The members of the joint commission considered numerous ideas over the next decade without any definitive decision to proceed. In 1913, the state of New York changed the name of its commission to the New York State Bridge and Tunnel Commission in recognition that a tunnel might be the best and most cost-effective solution. While the joint commission turned its attention to building a tunnel, powerful voices continued to advocate for a bridge in both engineering circles and in the popular press. As always with major transportation projects in the port of New York, the question of financing remained paramount. The never-ending argument over private capital versus public financing continued.

Traffic on the ferries across the Hudson increased even as the joint commission debated whether to proceed with a bridge or tunnel. A 1921 report documented the continued crush of traffic crossing the Hudson. The study divided the fifteen ferry lines into three groups: the "five-ferry group" running between lower Manhattan and Jersey City and Hoboken, a "six-ferry group" farther north to 42nd Street, and four additional ferry lines north of 42nd Street (table 6.1). Traffic on the two major ferry groups increased every year from 1914 to 1920 (table 6.2).

By 1914 the motor vehicle revolution was well under way. Henry Ford had produced his first Model T in 1908 and had sold the car for $825; millions of model Ts followed. Other automobile companies followed Ford's lead, and soon numerous manufacturers were producing simple and relatively inexpensive motor trucks to haul freight.

Despite the growing popularity of the internal combustion engine, the horse still played a prominent role in the port of New York's freight system as late as the 1920s. In 1914, less than 20 percent of the freight crossing the Hudson came via motor truck. Instead, an average of 7,655 horse-drawn drays hauled freight daily to lower Manhattan on the five-ferry group, and an additional 5,346 drays crossed the Hudson via the six-ferry group. Soon the truck began to dominate freight hauling. Motor trucks dramatically increased their share of traffic on the ferries from 1914 to 1920.

Bridges versus Tunnels

Long before New York and New Jersey formed their study commissions in 1906, engineers and railroad executives had put forth numerous plans to build bridges across the Hudson River. Gustav Lindenthal stands among the most prominent engineers advocating for a bridge across the Hudson rather than subaqueous tunnels. Born in 1850 in the part of the Austro-Hungarian Empire later to become Czechoslovakia, Lindenthal, like Wilgus, never completed an engineering degree, learning his engineering on the job. He emigrated to the United States and embarked on an engineering career as meteoric as that of William J. Wilgus. Henry Petroski devoted a long chapter to Lindenthal's career and achievements in his *Engineers of Dreams*.[7] In New York City, Lindenthal designed and built the Hell Gate Bridge over the East River for the New York Connecting Railroad. He later served as commissioner of bridges for the city of New York in 1902 and 1903 and oversaw the planning for both the Manhattan Bridge and Blackwell's Island Bridge, later renamed the 59th Street Bridge, over the East River.

In 1885, Lindenthal and the president of the Pennsylvania Railroad, Samuel Rae, announced plans for a railroad bridge over the Hudson to carry the trains of the Pennsylvania into Manhattan, twenty years before the Pennsylvania embarked on building a rail tunnel instead. Lindenthal believed it possible to build a suspension bridge across the Hudson of sufficient height over the river to allay any fears of the U.S. War Department and Navy. With a center span 2,850 long, the proposed bridge would rise 145 feet above the mean high tide of the river below, providing clearance for the tallest ships. The length of the bridge would exceed that of the Brooklyn Bridge, at the time the longest bridge in the world.

But even the wealthy Pennsylvania Railroad could not afford to build the bridge alone. In 1887, the Pennsylvania formed the North River Bridge Company to finance the bridge with the participation of the other New Jersey railroads. The plan included a massive union terminal in Manhattan for the passenger trains of all of the railroads. Once again, trying to get the competing private railroad companies to cooperate proved impossible, and the plans did not move forward. Over the next decade, Lindenthal often returned to his idea for a suspension bridge over the Hudson. In the very first issue of *Engineering News* in 1888, a lead story detailed Lindenthal's latest plans for a Hudson River bridge, praising the renowned engineer's brilliance: "There is probably no one on either side of the ocean who could be counted on more confidently to deal successfully with the intricate engineering problems than Mr. Lindenthal."[8]

When the New York State Interstate Bridge Commission began work in 1906, Lindenthal continued to argue for a bridge over the Hudson rather than tunnels. Writing to the *New York Times* in 1912, he admitted that the high cost, "ranging from $70 million to $100 million (including the costly right of way)," remained the major obstacle. He recognized that any solution had to accommodate the new auto age: "There is great public demand for a bridge particularly since the automobile traffic came into being. Automobiles can now cross the North River [Hudson] only on ferries requiring from twenty to forty minutes while on a bridge they could cross in three minutes." Lindenthal then explained why tunnels would not work: "New York is now virtually cut off from the continent as regards to all street and road traffic, including automobile and motor-truck traffic. Long tunnels are useless for them because the poisonous gases could not be removed by ventilation to prevent suffocation."[9] He firmly believed a bridge provided the only viable means for vehicular traffic to cross the Hudson. Without solving the problem of ventilating a tunnel, the carbon monoxide spewed out by internal combustion engines would create a death trap.

Lindenthal was still advocating a Hudson River bridge into the 1920s. Even in his seventies, at the end of his brilliant career, he proposed the construction of a massive bridge at 57th Street that would dwarf all of his earlier plans. He planned two colossal suspension towers taller than the Woolworth Building, then the tallest building in the world. With a width of 235 feet, the bridge included two decks—a top deck to accommodate ten lines of traffic and a lower level with eight railroad tracks:

Lower Deck:
4 Railroad tracks for freight
4 Railroad tracks for passenger trains from 7 railroad systems, with together
    24 tracks to the Union Station
2 Tracks for moving (or conveyor) platforms from under 57th Street

Upper Deck:
2 Tracks for rapid transit trains to the 9th Avenue elevated
2 Trolley tracks for surface cars
6 Lines of vehicular traffic
2 Sidewalks[10]

Lindenthal's complex plan included space for multiple modes of transportation: freight and passenger railroad traffic, a link to the 9th Avenue elevated railroad, and trolley tracks to connect the street railways on the New Jersey side to the extensive trolley system in Manhattan. Recognizing the advance of the auto age, he included six lanes for vehicles, three in each direction. Finally, sidewalks on each side of the upper level would provide spectacular views for people walking across the bridge.

Optimistically, the *New York Times* reported the "Great Hudson Span Close to Reality" and estimated the bridge to have a cost of $200 million, with a capacity to move 500,000 passengers, 12,000 vehicles, and 40,000 tons of freight an hour, and construction of the bridge to be completed in less than ten years. At a time in New York's history when complex projects such as Grand Central, Pennsylvania Station, the subways, and the Hell Gate Bridge were being completed, faith in the abilities of the leading engineers remained steadfast. The *Times* reported: "Gustav Lindenthal, the engineer who is directing the plan, had said a bridge can be built anywhere . . . that in engineering nothing is impossible, and that throwing of strands across the Hudson involves only methods that are known and have been tried."[11] Any transportation challenge, even the complicated needs of New York, could be solved if the necessary engineering talent, careful planning, and financial resources could be marshaled.

Lindenthal never secured the financing or the needed support from all of the railroads or the approval of the War Department for his 57th Street bridge. He lived to see the completion of the George Washington Bridge, 120 blocks to the north in 1933.

While Lindenthal advocated for a bridge, other engineers debated the comparative merits of a bridge versus additional subaqueous tunnels under the Hudson. In 1921, J. A. L. Waddell, an engineer and supporter of Lindenthal's 57th Street bridge, published a paper in *Transactions*, the prestigious and influential journal of the American Society of Civil Engineers.[12] In his professional work, Waddell focused on the costs and economics of long-span suspension bridges, and he had just published the *Economics of Bridgework*.[13] In the *Transactions* paper, he compared the costs and efficiencies of a bridge versus tunnels for crossing the "North River" (as many referred to the Hudson River).

Waddell pointed out that a tunnel under the Hudson would run approximately 90 feet underground, while a bridge would rise 180 feet above the river. The expenditure of energy needed to move either passengers or freight favored the tunnel. However, just as Lindenthal pointed out, a tunnel built for vehicular traffic as opposed to electric-powered trains posed a significant engineering challenge: "The safe ventilation of a tube carrying automobile traffic is a yet unsolved problem. . . . Ventilation, if feasible would be exceedingly expensive, and the velocity of the passing air would be excessive." Waddell warned that if traffic broke down, the carbon monoxide from the autos and trucks in the tunnel "might result in a holocaust," a strong warning indeed.[14] He added that he held no objection to building a tunnel per se. If it failed for automobile traffic, electric-powered railroads or a moving conveyor could be employed to load cars and trucks and move them under the river.

After explaining in detail the basis for his cost calculations, including the cost of suspension wire for the bridge and the cast iron rings used to line a subaqueous tunnel, he compared the costs for five alternatives, starting with a highway suspension bridge spanning at least 2,900 feet:

Highway bridge . . . . $ 32,500,000
Four-lane highway tunnel . . . . $ 49,000,000
Electric-railway bridge . . . . $ 34,500,000
Four double-track electric railway tunnels . . . . $ 48,000,000
Eight single-track electric railway tunnels . . . . $ 30,000,000[15]

Even though no long, subaqueous tunnel for vehicular traffic had yet been built anywhere in the world, Waddell's estimate included elaborate calculations for the costs of a ventilation system sufficient to remove exhaust even in the event of complete blockage of traffic to prevent a "holocaust." The cost advantage of the bridge over a tunnel exceeded $16 million, one-half of the total projected costs. Waddell did point out that the combination of a bridge for vehicular traffic and two single-track tunnels for freight would cost $40 million, but that was still significantly less than an all-vehicle tunnel.

According to Waddell, the most efficient and cost-effective means to move freight in the port of New York remained a tunnel for electric-powered trains. Electric trains eliminated the need for an expensive ventilation system. More than a decade earlier, Wilgus's small-car freight system utilizing electric power offered precisely the solution Waddell estimated to be the least expensive!

Planning for the Holland Tunnel

While the debate over a bridge to cross the Hudson continued, the bridge and tunnel commissions for New York and New Jersey finally recommended the construction of a subaqueous tunnel for vehicular traffic under the Hudson between Jersey City and lower Manhattan at Canal Street. With a planned completion date of 1924, the new tunnel would create, in the words of an *Engineering News-Record* editorial, "a highway under the Hudson" and usher in another revolution in American transportation. A tunnel would finally solve the centuries-old problem of New York's isolation from the mainland—"a business community of six million people, cut off from its outliers and from the whole country beyond by a broad shipping channel."[16]

The joint commission of the two states, after almost two decades of investigations, reports, and proposals, selected a preliminary design proposed by the prominent engineer Clifford M. Holland. The commission also asked Holland to serve as the chief engineer for the project. Both states passed enabling legislation in 1919. New York's legislation specified the project's scope: "To construct . . . a tunnel or tunnels under the Hudson River between a point in the vicinity of Canal Street on the island of Manhattan and a point in Jersey City, in the state of New Jersey, for the exclusive use of pedestrians and of vehicles not operated by public service corporations."[17] The reference to "public service corporations" singled out the railroads, prohibiting the use of the planned tunnel by the private railroads.

New York and New Jersey agreed to fund the cost of construction equally. Once again, a momentous transportation initiative required the commitment of public funding to move forward. Private capital was unwilling to undertake such an expensive and daring project—the construction of the longest subaqueous tunnel in the world and the first dedicated solely for vehicular traffic. Whether the railroads realized it or not, the tunnel signaled the end of the dominance of the railroads in the port of New York and further confirmed the ascendency of the highway and the automobile.

By the time the Holland Tunnel project began, the tunnels of two railroads— those of the Pennsylvania Railroad and of the Hudson & Manhattan—already connected New Jersey to Manhattan. Under the East River and the Harlem River, ten subway tunnels provided mass transit links to Brooklyn, Queens, and the Bronx. When the Brooklyn Bridge opened in 1883, cable cars moved passengers across the span. Both the Williamsburg Bridge (1903) and the Manhattan Bridge (1910) over the East River combined pedestrian walkways, subway tracks, and vehicular traffic. All three of the East River bridges combined different modes of transportation. Earlier plans for the tunnels under the Hudson proposed both rail and vehicular traffic in double-decked tunnels. But a double-deck tunnel required a much larger diameter. Since tunnel costs increased dramatically with the diameter of the tunnel, a two-level tunnel proved to be prohibitively expensive. In 1920, the commission decided to construct a single-deck tunnel to accommodate only one mode of transportation—cars and trucks powered by the internal combustion engine.

For the "Hudson River Vehicular Tunnel Project," the joint commission of the two states first established a board of consulting engineers to serve in a key advisory capacity. Such an advisory commission was standard for all major engineering projects. Almost twenty years earlier, when the Grand Central project began, Wilgus had established the Electric Traction Commission to advise on all major engineering questions involved in the new technology on which the entire project rested—electrification. Following this tried and true model, the joint commission recruited a group of outstanding engineers and asked Wilgus not only to join the board but also to serve as its chairman, a singular honor and a clear recognition of his stature in the engineering profession. In his autobiography, Wilgus referred to the invitation as "one of my most notable engagements." He wrote that he felt honored to be in the company of a group of eminent engineers that included "Professor William H. Burr of Panama Canal fame . . . and J. Vipond Davies, who in previous years had been engaged in the world famous creation of the Hudson & Manhattan subaqueous tunnels between Manhattan and New Jersey."[18]

On May 15, 1919, Wilgus attended a meeting of the New York–New Jersey Bridge and Tunnel Commission to discuss his appointment to the Board of Consulting Engineers. At the meeting the commission defined the key role of the consulting engineers:

- It shall be the duty of the Board of Consulting Engineers to advise the New York and New Jersey Commissions and the Chief Engineer, to follow the progress of the work, whether preliminary or constructive, and to formally report thereon to the Commissioners every third month.
- Hold regular meetings bi-weekly until the type of tunnel to be constructed and the design thereof has been determined upon.
- Assemble at 24 hour notice and shall consider such problems relative to the project as may be presented . . . and give its advice thereon.[19]

The responsibilities of the Board of Consulting Engineers included deciding on the final design of the tunnel and supervising initial construction. If problems arose, the members of the board pledged to meet within twenty-four hours to deal with any engineering issues.

Wilgus agreed to serve on the board at a salary of $10,000 per annum, with each state commission paying half. This was a substantial consulting fee, almost equal to the salary of the chief engineer, Clifford Holland, who was paid $12,500 and worked full time on the tunnel project. Although the board's bi-weekly meetings required a major commitment, Wilgus also continued his other consulting work.

The Board of Consulting Engineers was officially formed on June 26, 1919, when the joint commission passed a resolution appointing Wilgus and the other members to the board. At the board's first meeting in October, his fellow engineers confirmed the commission's decision by electing Wilgus the chairman. Now officially constituted, the Board of Consulting Engineers turned to the tasks that faced them. Among these were three major considerations: the volume of traffic that would be assumed to use the new tunnel, what method should be used to construct the tunnel, and the design of the tunnel's ventilation system.

The board first focused on analyzing current traffic crossing the Hudson to develop projections for the volume of traffic the tunnel would carry. Traffic projections formed a crucial part of the planning process because the plan called for the tolls paid by the vehicles to finance the project. Then two critical engineering questions needed to be decided: the type of construction to be used and, particularly crucial, the design of the ventilation system for a tunnel to be filled with vehicles powered by the internal combustion engine. Chief Engineer Holland organized a support staff to work with the consulting engineers to decide all important engineering questions, complete all detailed engineering calculations, draw construction plans, and solicit bids.

A Study of Hudson River Traffic

To develop projections of the traffic to use the new tunnel, Wilgus directed Holland's staff to conduct a detailed study of current traffic crossing the Hudson by ferry. The staff collected passenger and vehicular information for ferry operations for the time period from 1914 to 1920. In the summer and early fall of 1920, the staff carried out traffic surveys on the ferries, recording the types of vehicles crossing between New Jersey and Manhattan. The staff also interviewed drivers and asked their origin and destination.

For the "five-ferry group" serving lower Manhattan (see table 6.1), traffic eastward averaged 5,836 vehicles during weekdays. Horse-drawn vehicles accounted for 2,856 (48.9%) of the total; motor trucks, 1,864; and the remainder—"pleasure cars"—1,116. Most of the traffic—4,284 vehicles (73% of the total)—originated in Hudson County, New Jersey, with an additional 1,355 vehicles from either Essex or Union Counties. Fewer than 200 vehicles came from as far away as either Bergen or Passaic Counties.[20] Most of the traffic from beyond Hudson County consisted of private automobiles. Traffic back across the Hudson to New Jersey averaged 6,006 vehicles each weekday, and most of the volume consisted of local traffic.

For the "six-ferry group" farther north, the number of vehicles crossing to Manhattan averaged 3,893 each day. If a significant number of the vehicles crossing by ferry switched to the tunnel and paid a $0.50 toll, the two states would share more than sufficient revenue to pay the bonds borrowed for construction costs. The governors and politicians of both states agreed to support the project on the premise of a "self-financing" tunnel without the need to use general state revenue for either construction financing or ongoing maintenance.

The traffic forecast model assumed that 20 percent of the freight currently moved across the harbor by lighters would be switched to motor trucks using the new tunnel. For each ferry group, the engineers provided a separate estimate of the percentage of current traffic to use the new tunnel: for the five-ferry group, 62 percent, and for the six-ferry group farther up the Hudson, only 14 percent (table 6.3). With over ten million vehicles projected to use the tunnel in the first year of operation, the model estimated revenue of over $5 million.

In addition to traffic projections, the engineers needed to determine the capacity of each lane to be able to recommend how large a tunnel to construct. To arrive at reasonable estimates, the staff studied traffic moving across the Brooklyn, Manhattan, and Williamsburg Bridges over the East River. In August and Sep-

Table 6.3. Board of Consulting Engineers' forecast of one-way traffic in the Holland Tunnel, 1924–1944 (number of vehicles)

Year	Five-ferry group (62% to tunnel)	Six-ferry group (14% to tunnel)	Lighters (20% to tunnel)[a]	Avg. daily no.	Total no. for 340 days
1924	9,900	900	5,000	16,500	5,610,000
1930	20,300	1,600	5,600	28,000	9,250,000
1935	31,000	2,600	6,200	40,500	13,800,000
1940	44,500	3,800	6,800	55,400	18,800,000
1945	56,600	4,900	7,400	69,100	23,500,000

SOURCE: "Report of Board of Consulting Engineers, Jan. 24, 1920," William J. Wilgus Papers, Manuscripts Division, New York Public Library, box 49.
[a] Assuming total traffic in four-ton truck units.

Table 6.4. Estimates of traffic capacity for Holland tunnel

Number of lanes and type of traffic	Hourly	Daily	Annual
One lane, including horse-drawn vehicles	480	5,850	1,990,000
Two lanes, with horse-drawn in one lane	1,600	19,500	6,635,000
Three lanes, all motor-driven vehicles	3,020	36,800	12,500,000

SOURCE: State of New York, *Report of the New York State Bridge and Tunnel Commission*, 20.

tember of 1919, a strike by the workers of the Brooklyn Rapid Transit Company (BRT) created higher-than-normal traffic on the bridges. The strike provided the opportunity to study capacity during a period of exceptionally high demand.

On August 30, 1919, traffic over the Williamsburg Bridge averaged 1,021 vehicles an hour in one direction. Two lanes of traffic provided a "fast" lane reserved for automobiles and motor trucks and a "slow" lane for horse-drawn drays and slower motor vehicles. On the Manhattan Bridge, which had three lanes in each direction, traffic averaged 2,681 vehicles an hour. A key reason accounted for the higher capacity on the Manhattan Bridge: regulations prohibited horse-drawn vehicles on the bridge. Eliminating horse-drawn drays dramatically increased traffic capacity. The traffic study concluded with three separate estimates of capacity through the tunnel in one direction (table 6.4).

Clearly, if traffic through the Holland Tunnel reached an estimated five million vehicles each way in 1924, the tunnel needed at least two roadways in each direction. If there were only one roadway in each direction and horse-drawn drays were allowed to use the tunnel, the tunnel could not accommodate the projected traffic and would be obsolete from the day it opened. Unequivocally, Wilgus and the consulting engineers recommended "two twenty-foot roadways, each providing two lines of traffic . . . to meet traffic requirements between Canal Street, Manhattan and Twelfth Street, Jersey City," with no horse-drawn vehicles allowed.[21] The engineers also recommended that each of the two tunnel "tubes" accommodate two lanes of traffic moving in the same direction.

Almost immediately, controversy arose over the capacity of the planned tunnel. Critics, including Gustav Lindenthal, argued that with only two lanes in each direction, the tunnel would not have sufficient capacity to move the ever-increasing amount of vehicular traffic crossing the Hudson. Lindenthal, among others, demanded that the Board of Consulting Engineers plan for each tube to have three lanes of traffic in each direction.

The Bridge and Tunnel Commission held a series of public hearings in early 1920 to solicit comments on the draft plans. Several speakers voiced opinions that freight traffic by motor vehicles had grown so dramatically that "it will more than utilize the maximum capacity of these (two-way) tunnels in less than six months' time. . . . We should have three lane traffic and even it would reach its maximum in a year's time."[22] Lindenthal's concerns echoed among the public.

In response, Wilgus wrote a letter to Holland in which he reiterated his position on the tunnel's traffic capacity. He referred to "trucking interests" who viewed the tunnel as a "cure-all." They wanted to use the tunnel to move a huge volume of freight under the Hudson by motor truck. Wilgus, on the other hand, believed that the legislative mandate required the tunnel to "be in harmony with the surface capacity of the streets with which it is to connect at each end." He envisioned the tunnel as a link between the existing streets in Manhattan and Jersey City, not as a super highway under the Hudson and certainly "not a sub-stitute for existing means of handling freight."[23] He then mentioned the solution to moving freight across the harbor that he had proposed twelve years earlier: "It was proposed that freight should be moved in electrically operated multiple-unit trains." Wilgus never abandoned his plan for the freight subway system he had proposed to the Public Service Commission of New York in 1908 even as he served as the chief engineering consultant for the first vehicular tunnel under the Hudson.

Above all else, Wilgus warned Holland not to be persuaded to increase the capacity of the tunnel as Lindenthal and others demanded. In his mind, the solution to moving freight in the port of New York could not be solved by the motor truck. But while Wilgus continued to believe in a railroad solution for moving freight, the trucking companies viewed the tunnel as a golden opportunity to dramatically increase their business. With a vehicular tunnel, they could capture a greater share of freight hauling in the port. And if New Jersey also expanded its highway construction, then the trucking companies could challenge the rail-roads' monopoly of freight hauling over much longer distances than just between Jersey City and lower Manhattan.

Determining a Method for Tunnel Construction

A second major engineering question for the Board of Consulting Engineers in-volved recommending the method to be used to construct the tunnel. The board carefully considered two competing construction techniques: a trench tunnel and a shield-driven tunnel. To construct a trench tunnel, dredges dig out a trench along the bottom of a river. On shore, workers construct sections of the tunnel, which are towed into position and lowered under water into the trench. Divers connect the tunnel sections together; then, compressed air is used to force out all water to enable workers to seal the connections between sections, making the tunnel watertight.

Wilgus brought real expertise to this dilemma. In his early career, not only had he achieved fame for his work at Grand Central, but he had also received praise for his design and supervision of the Michigan Central Railroad's Detroit River Tunnel. Wilgus began work on the project in 1906, while still employed by the New York Central Railroad, the parent company of the Michigan Central. Be-tween 1906 and 1910, Wilgus supervised the planning and completion of the tunnel which required a unique construction technique. At first, plans called for a shield-driven tunnel, but test borings of the riverbed found very liquid condi-tions with poisonous gases trapped in the silt.

Not the first time in his career, he claimed credit for a revolutionary idea: tunnel construction utilizing the "trench and tremie method." He explained in some detail how this idea came to him:

> On my journey from Detroit to New York . . . I gave free reign to my imagination in a series of steps: (1) had the river bed only been impermeable solid rock instead of clay, how easy it would be to bore out tunnels; (2) that not having been provided by nature, then why not create artificial solid rock of concrete in a dredged trench through which holes might be bored with safety in the dry; (3) but hold—if the latter should be feasible, why not insert forms in the artificial rock, of inside dimensions sufficient to permit the placing of interior lining, thus making the interior boring unnecessary and save expense?[24]

Wilgus's first thought was to dig a trench in the bottom of the Detroit River and fill the trench with concrete. A tremie consists of a long tub through which waterproof concrete can be pumped under water. The lower end remains immersed in the concrete, and as the pumping continues, the concrete expands and continues to cover the lower end of the tremie as more concrete arrives. Engineers and builders still use the tremie system for underwater concrete construction.

In a flash of genius Wilgus decided to first lower two steel tubes into the trench under the river, side by side. Once divers secured the tubes in place, the builders used tremies to pump concrete under water to surround the two steel tubes. Each tube had pins and sockets at the ends to connect with the adjacent tube; divers then locked the tubes together and secured the joints with rubber gaskets.[25] Each steel tube measured 23 feet 4 inches in diameter and 262 feet in length. The tunnel required a total of ten steel tubes to create the interior space of the tunnel. Once compressed air forced out the water, workers connected the sections and lined the tunnel with a concrete inner wall, covering the steel tubes for added strength.

The Detroit River Tunnel totaled 12,792 feet from portal to portal, including underground approaches to the underwater sections on each side of the river. The *Railway Age Gazette* listed the total cost as $4,775,306, with just about half spent on the subaqueous section in the trench under the river.[26] When the tunnel was completed, a familiar controversy broke out over the electrification. Electric-powered engines used in the tunnel eliminated the need for an elaborate ventilation system, as they had eliminated the smoke and gases in the Park Avenue Tunnel. Just as for the Grand Central electrification, Wilgus insisted on direct current and a third rail to provide current to the electric engines.

Justly proud of his work on the Detroit Tunnel, Wilgus explained that he continued to work on the tunnel after he resigned from the New York Central after the Woodlawn Wreck because New York Central president Newman and William K. Vanderbilt (grandson of Commodore Vanderbilt) "requested me to see it through to the end."[27] It certainly provided him with sweet satisfaction to be asked by the president and chairman of the board of the Central to stay on after the bitterness he felt over the Woodlawn Wreck controversy.

Confident in the truly innovative nature of his trench and tremie method, Wilgus, along with H. A. Carson—another of the consulting engineers, who had achieved fame building Boston's subway and East Boston tunnel—applied for and received a patent for his innovative subaqueous tunneling techniques. Both Carson and Wilgus agreed to allow the tunnel company and the Michigan Central to use their patented construction technique without a licensing fee.

Wilgus received a laudatory letter from the president of the Detroit River Tunnel Company in 1910 after the official completion of the tunnel: "The tunnel, as now constructed, is a great advance over any former means of subaqueous construction, and as it was largely due to your efforts that the method followed in the construction of the tunnel was adopted, in the face of opinions of able engineers that the same would not be a success."[28] Once again in his career, William Wilgus stood by his convictions and advocated a radically new construction technique, placing his engineering reputation on the line, just as he had when he proposed the massive Grand Central project. In his memoirs he triumphantly concluded, "Success, not failure, attended that for which I risked so much."[29]

Ten years after his success with the Detroit River Tunnel, Wilgus was now chairing the Board of Consulting Engineers, which planned to build the longest subaqueous vehicular tunnel in the world. Their first major engineering decision involved deciding on the method to employ for building the tunnel: either the trench method that Wilgus had employed in Detroit or the subaqueous shield method.

The consulting engineers noted that the amount of traffic on a waterway would strongly affect the cost of using the trench method. To dig the trench, barges would have to be anchored in one spot for extended periods. While this system had worked in Detroit, boat traffic on the Hudson vastly exceeded traffic on the Detroit River. To illustrate, the staff conducted a count of vessels, both north- and south-bound, passing the pier at Canal Street for a twenty-four hour period on December 18, 1920. This pier was situated where plans called for the tunnel to cross under the Hudson and where numerous dredges, barges, and construction vessels would anchor for a year or more while the trench excavation proceeded. A total of 1,574 vessels passed in this twenty-four-hour period, illustrating the scale of waterborne traffic in the port of New York (table 6.5). Tugs, barges, lighters, and car floats associated with hauling freight across the Hudson accounted for over half of all marine

Table 6.5. Hudson River marine traffic passing Canal Street pier, December 18, 1920

Class	No. of boats	% of total traffic
Tugs and lighters		
Tug boats	402	26
Tugs with barge	180	11
Tugs with lighter	126	8
Tugs with car float	116	8
Lighters under own power	71	5
Subtotal	895	58
Ferry boats	512	32
Small power boats	84	5
Ocean-going boats	49	3
Miscellaneous	34	2
Total	1,574	100

SOURCE: State of New York, *Report of the New York State Bridge and Tunnel Commission*, plate 11.

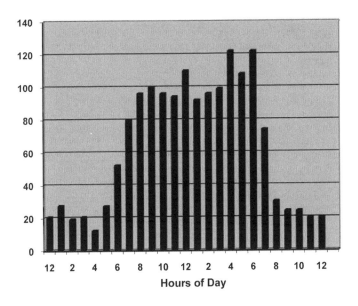

140

120

100

80

60

40

20

0

12 2 4 6 8 10 12 2 4 6 8 10 12

Hours of Day

Number of north-
and southbound
vessels on the
Hudson River,
December 18, 1920.

traffic on the Hudson (58%); ferry boats made up most of the rest of the traffic. An average of 65 vessels of all types passed the Canal Street pier each hour.

Each day thousands of vessels moved up and down the Hudson River and filled all of the waterways in the port of New York, as they had for centuries. The busiest port in the United States and the world simply could not afford to have a major construction project block one of its main waterways, the Hudson River. While water traffic lessened during the overnight hours, the Hudson remained filled with boat traffic almost all hours of the day.

Colonel Edward Burr, the U.S. engineering officer in charge of the New York District, discussed the potential hazards on the Hudson River before the American Society of Civil Engineers in October 1920. He elaborated on the dangers posed by dredging and construction: "Every dredge or other machine working in a fixed position in a channel is an obstruction to traffic, which is greatly increased if moorings are necessary, as in swift tidal currents, and the congestion existing in the main channels of New York harbor . . . creates a most serious condition."[30]

The nature of the bottom material in the Hudson posed another serious problem for utilizing the trench method. In Detroit, the river bottom consisted of clay, relatively easy to excavate. Under the Hudson, soft silt required extensive shoring of the walls of the trench, a time-consuming and expensive operation.

Both the consulting engineers and Holland's engineering staff reached the decision that the shield method offered the most practical and safest method to build the tunnel. Both the Manhattan & Hudson and the Pennsylvania railroads had employed the shield method to construct their railroad tunnels under the Hudson. Constructing the Holland Tunnel with a shield method offered other advantages. In Jersey City, the Erie Railroad owned the waterfront property needed for the tunnel approach. Never cooperative, the Erie insisted that tunnel construction not interrupt its train operations; if it did, the railroad threatened

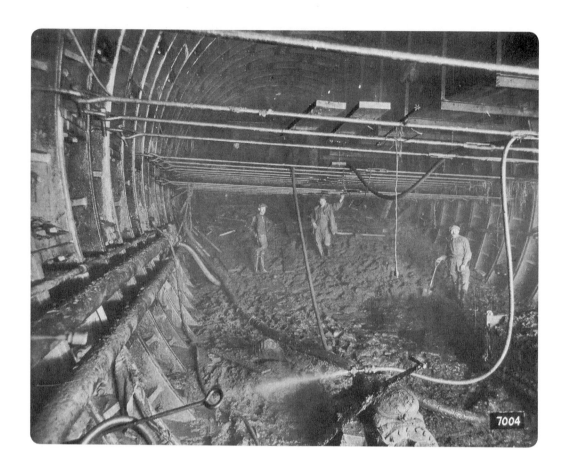

Construction
of the Holland
Tunnel, 1924.
Note the dramat-
ically increased
size of the
shield used
for the Holland
Tunnel versus
much smaller
shield used
for the Hudson
& Manhattan
Railroad tunnels
in 1908 (second
illustration in
this chapter),
today's PATH
tunnels.

to sue to stop construction. The shield method would allow the tunnel to pass under the Erie rail yard and not disrupt the train yard on the surface above.

In their recommendation to the Bridge and Tunnel Commission, the Board of Consulting Engineers summarized their argument for shield tunneling: "It is plainly evident from the width of the river, the great volume of river traffic, waterfront conditions and conditions of the riverbed that . . . the Hudson River is most suited to shield tunneling."[31] Once again, Wilgus had led an engineering team that carefully marshaled data, systematically compared alternative construction techniques, engaged in open and frank debate of each method, and then reached a careful, measured recommendation. Clifford Holland and the New York–New Jersey joint commission adopted the recommendation of the consulting engineers and proceeded to develop detailed plans for a shield tunnel, with twin cast iron tubes, each to have two lanes of traffic. They estimated the total cost of the tunnel at $28,669,000.

Throughout the spring of 1920, Holland and his staff worked furiously on detailed plans and specifications in order to put the tunnel out to bid. The staff and the consulting engineers devoted a great deal of attention to the design of the cast iron tubes that would form the walls of the tunnel. Since the actual conditions to be encountered as the shield moved forward could not be determined before construction, the engineers designed the cast-iron tunnel rings to withstand enormous external pressure. In the language of the engineering profession, the analysis "comprised the determination of true equilibrium polygons for different conditions of loading whence the resultant thrusts and bending moments in the tunnel rings were completely determined." In his 1921 report to the joint commission, Holland and his staff included a fifty-five-page appendix filled with calculations and charts detailing the careful analyses carried out to determine the exact specifications for the cast iron rings.[32] Wilgus and the other consulting engineers signed off on all of the calculations.

Initially, the engineers estimated four years for the construction of the tunnel. As work progressed, Holland and his staff revised the construction schedule, trying to keep the work to a four-year timetable. As with all major building projects, especially one as complicated as constructing the longest vehicular tunnel in the world, keeping on schedule proved to be a challenge. In December 1920, the staff once again revised the schedule. Just as Wilgus had divided the immensely complicated Grand Central excavation of the train yard into twelve separate "bites," Holland divided the work into nine separate contracts, with work proceeding on a number of them simultaneously (table 6.6). One crucial example illustrates the precision necessary: tunneling from both the New York and New Jersey sides of the Hudson began at the same time. Very careful survey work ensured that the

Installation of cast iron tunnel walls in the Holland Tunnel, 1923.

5474

Table 6.6. Holland Tunnel timetable established in December 1920

Contract	Plans and specifications ready	Contract delivered	Contract competed
1. New York land shafts	Aug. 16, 1920	Oct. 1, 1920	Oct. 1, 1921
2. New Jersey land shafts	Jan. 1, 1921	Mar. 15, 1921	Jan. 15, 1922
3. New York river tunnel	May 15, 1921	July 1, 1921	July 1, 1924
4. New Jersey river tunnel	May 15, 1921	July 1, 1921	July 1, 1924
5. New York approach	Nov. 15, 1921	Jan. 1, 1922	July 1, 1924
6. New Jersey approach	Nov. 15, 1922	July 1, 1922	July 1, 1924
7. Ventilation building, New York	Nov. 15, 1922	July 1, 1923	July 1, 1924
8. Ventilation building, New Jersey	Nov. 15, 1922	July 1, 1923	July 1, 1924
9. Installation of ventilating equipment	Nov. 15, 1923	July 1, 1923	Dec. 31, 1924

SOURCE: State of New York, *Report of the New York State Bridge and Tunnel Commission*, 36.

tunnel tubes met precisely under the Hudson River, joining the two separate sections, the one from the New Jersey side and the other from New York.

Resolving the Ventilation Issue

An absolutely essential part of the engineering for the Holland Tunnel involved the design of the ventilation system. In fact, the entire project depended on a ventilation system capable of removing the deadly carbon monoxide gases produced by the internal combustion engine. Without an exhaust system capable of removing the carbon monoxide the tunnel would be a death trap.

As he had before, Wilgus again led an effort to adopt a radical new technology to enable a large construction project. Twenty years earlier, the success of the entire Grand Central project had rested on the successful switch from steam power to electricity—at the time an unproved technology for moving heavy passenger trains. Electrification eliminated steam and smoke from the Park Avenue tunnel and the new two-story underground train yard. For the Holland Tunnel, removing the deadly gases from the tunnel and protecting the traveling public required similar innovation.

Determining the amount of gases produced and their potential impact on drivers and passengers demanded careful investigation in order to determine the scale of the ventilation system. There was little precedent for such an analysis. The Holland Tunnel was to be much longer than any other vehicular tunnel in the world. The two existing railroad tunnels under the Hudson did not require elaborate ventilation because the trains used electric engines, which produced little air pollution. Since no existing vehicular tunnel provided a model, the consulting engineers commissioned two panels of outside experts to conduct detailed experiments to determine the amount of carbon dioxide produced by cars and trucks and the risk to drivers and passengers from inhaling the exhaust gases.

The Board of Consulting Engineers engaged the U.S. Bureau of Mines to study the amount of gases produced by trucks and automobiles. The Bureau of Mines had expertise in the study of dangerous gases because miners often encountered gases, including carbon monoxide. At its experimental station in Pittsburgh, the bureau maintained the necessary equipment and staff to carry out the research on the effects of carbon monoxide.

A. C. Fieldner, the Bureau of Mines' supervising chemist in Pittsburgh, directed the research between December 1919 and September 1920. The research plan included testing vehicle emissions during both winter and summer months. An elaborate testing schedule included 101 cars and trucks, varying in horsepower from 18 to 40, with the trucks having a maximum carrying capacity of four and a half tons. Today, when both cars and trucks possess immense horsepower, the vehicles from the early 1920s seem quaint. At the dawn of the automobile age, the tunnel engineers could not have imagined the 5,000-pound SUV used to ride to the local convenience store or a 40,000-pound, 60-foot-long tractor trailer hurtling down the interstate at 70 miles an hour. In fact, an important research question in 1920 involved how much a four-ton truck with a 40-horsepower motor would slow down as it climbed the 3 percent grade upward to exit the tunnel.

A letter from the director of the Bureau of Mines pointed out the groundbreaking nature of the investigation: "The construction of the Vehicular Tunnel between New York and New Jersey under the Hudson River involves some problems in ventilation and design which have not heretofore been investigated. These problems arise from the substitution of automobiles and trucks for horse-drawn vehicles and the necessity of providing ventilation adequate for the removal of exhaust gases."[33] At Grand Central, Wilgus had directed a technology shift from steam to electricity, while at the Holland Tunnel, motive power shifted from the horse to the internal combustion engine.

The chemists at the Bureau of Mines designed an apparatus to sample exhaust gases. Carbon monoxide, the poisonous part of the exhaust, results from incomplete combustion of carbon in gasoline. To arrive at reasonable estimates, the testing included a variety of automobiles and trucks with different carburetor settings and different grades of gasoline. Tests over the summer found the average amount of carbon monoxide in exhaust gases to be 6.7 percent, but there was significant variation in these percentages, from just over 1 percent to over 13 percent (table 6.7).

Fieldner reported that the highest percentage of carbon monoxide resulted from the 3 percent grade at each end of the tunnel. With the test data, the engineers calculated the average carbon monoxide produced per hour to be between 70.7 and 199.3 cubic feet. With the amount of carbon monoxide produced and estimates of the number of vehicles using the tunnel per hour, the engineers calculated the amount of carbon monoxide the ventilation system needed to remove. Of course, the ventilation system specifications included a significant safety factor, with more capacity than needed under the most adverse conditions.

Estimating the amount of carbon monoxide to be removed constituted only the first step. A second research team led by Dr. Yandell Henderson of Yale Uni-

Table 6.7. Carbon monoxide emissions from representative makes
of passenger cars and trucks in 1919–1920 tests

Car no.	Type of vehicle	Speed (mph)	Miles per gallon	% combustion	% carbon monoxide in exhaust
10	5-passenger	15	11.16	61	1.7
84	½-ton truck	15	15.39	90	1.7
38	3½-ton truck	10	6.55	87	1.9
1	5-passenger	15	27.3	100	3.7
9	5-passenger	15	13.26	84	3.7
11	7-passenger	15	18.61	93	9.3
57	3½-ton truck	10	4.81	65	10.6
76	¾-ton truck	15	10.66	59	10.7
44	5-passenger	15	10.26	49	13.2

SOURCE: State of New York, *Report of the New York State Bridge and Tunnel Commission*, 134.

versity studied the effect of carbon monoxide on people. No systematic research on the psychological effects of carbon monoxide on people had ever been carried out. The principal research questions included

- the maximum amount of carbon monoxide that could be absorbed without producing discomfort;
- the rate at which the driver of a vehicle would absorb carbon monoxide; and
- investigation of the possibility of individual variations in susceptibility to carbon monoxide poisoning.

Yandell and a team of graduate students working at the Bureau of Mines Experiment Station at Yale experimented on themselves in a gas-tight chamber for one person of 226 cubic feet capacity. Each test lasted one hour, and the amount of carbon monoxide admitted to the chamber varied from two to ten parts per thousand (table 6.8). The researchers chose one hour for each test because traveling from one end of the Holland Tunnel to the other at 3 miles per hour would take 31.4 minutes. In addition to the individual tests, the research team used a larger chamber of 12,000 cubic feet, with an automobile engine running inside, to accommodate groups of test subjects.

The research report sent to the tunnel commission included a diagram illustrating a test subject sitting inside the smaller, individual chamber. Before entering the chamber, each subject gave a blood sample and then gave a second one leaving the chamber after the hour-long experiment. The most common effects proved to be "the typical carbon monoxide, or oxygen deficiency, headache, . . . a distinctly localized pain, usually frontal, throbbing, . . . sometimes accompanied by more or less nausea, readily increasing to vomiting."[34]

It is difficult to imagine that these experiments on human subjects could be replicated today without serious questions raised by a university institutional review board, or IRB. All universities that carry out any research with human subjects must have a federally mandated IRB to review proposed research. The goal of the IRB is to ensure that human subjects are not at risk, with an overarching principle: "do no harm."

After an extensive number of tests, the researchers summarized their findings and compared blood saturation with carbon monoxide to an "equilibrium value" that represented the maximum amount of carbon monoxide blood can absorb from unlimited exposure. The research report concluded that for an exposure of 45 minutes, four parts per ten thousand of carbon monoxide and up to six parts per thousand "would afford complete safety, but also comfort and freedom from disagreeable effects."[35]

The research on exhaust and the effect of carbon monoxide provided Wilgus and the other consulting engineers with all the data needed to design the ventilation system. Drawing on the studies of existing ferry traffic, they estimated that with two lanes of traffic each tunnel tube had a capacity for an average of 2,113 vehicles per hour. The engineers added an 11 percent increase to the estimated traffic for safety purposes. They then ran calculations to determine the amount of fresh air required per minute. The amount of fresh air needed required forty complete changes of air in the tunnel each hour.

The engineers evaluated two ways of providing ventilation: the longitudinal method and the distributive method. In the longitudinal method, used in railroad tunnels, air is forced into the tunnel at one end and exits at the other. For the Holland Tunnel, given its length, the air would move at high velocity. Estimates of air velocity exceeded 50 miles an hour, far too high for comfortable travel. In addition, the energy needed to drive the fans would be excessive. Thus, the distributive method of ventilation was chosen.

To implement a distributive ventilation system, the engineers designed two

Table 6.8. Experimental results for carbon monoxide tolerance by test subjects

CO in test chamber (parts per 10,000)	Equilibrium blood saturation[a] (%)	Blood saturation after 1 hour (%)	Headache
2	23	11–12	None
4	36	16–20	None
6	47	18–26	None or slight
8	53	32–34	Distinct
10	58	38	Marked

SOURCE: State of New York, *Report of the New York State Bridge and Tunnel Commission*, 31.

[a] Blood saturation level if the person were subjected to the concentration of carbon monoxide indefinitely.

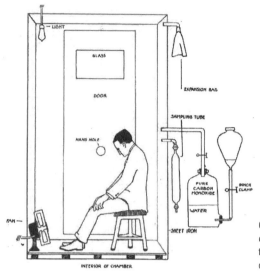

LIGHT

GLASS

EXPANSION BAG

DOOR

SAMPLING TUBE

HAND HOLE

PURE CARBON MONOXIDE

PINCH CLAMP

WATER

FAN

SHEET IRON

INTERIOR OF CHAMBER

Carbon monoxide testing chamber, 1922: chamber used at Yale University to test the effects of carbon monoxide as part of the planning for the Holland Tunnel.

air ducts for the circular tunnel, one under the roadway to supply fresh air and the other over the roadway to draw off the exhaust gases. Air moving in both ducts would be controlled by adjustable vents. The engineers planned two ventilating shafts, one on each side of the Hudson, built just off the shoreline, to draw in fresh air and vent exhaust gases. On each shore, two buildings housed ventilation fans powered by electricity. To ensure that the ventilation removed carbon monoxide to a safe level, the consulting engineers commissioned engineers at the University of Illinois and the Bureau of Mines to conduct further experiments on the ventilating system before completion of the final construction drawings.

This additional research found the original plans for the ventilation equipment to be inadequate, and Wilgus and the consulting engineers recommended major changes. New specifications called for the tunnel diameter to be increased to 29 feet 6 inches to accommodate larger ducts for both fresh air and exhaust gases.[36] The board also insisted on auxiliary batteries to power the fans in the event of a power failure. These changes contributed substantially to the rising costs for the tunnel and also required longer construction time. First estimates prepared in 1919 placed the total cost for building the tunnel at $28.7 million. In a detailed report written in 1923, Wilgus estimated the new cost to exceed $41 million, driven in part by the need to increase the capacity of the ventilation system.[37] Of course, the engineers' recommendations met with resistance because of the higher costs and longer timetable. Wilgus refused to compromise, and the other consulting engineers unanimously supported Wilgus's position. In the face of their insistence on the proposed design changes, the Bridge and Tunnel Commission had no choice but to proceed with the modified tunnel design with larger air and exhaust ducts and larger ventilation equipment. The modifications added millions of dollars to the final costs and delayed completion of the tunnel but ensured the safety of the traveling public.

Above: Mock-up of expanded Holland Tunnel (diameter of 29 feet 6 inches) to allow for adequate ventilation, 1923. The mock-up shows ventilation spaces below the roadway for fresh air and above the roadway for exhaust. Silhouettes are of a 1920s automobile and truck.

Left: Construction of ventilation ducts above and below the roadway in the Holland Tunnel, 1925.

The Construction of the Holland Tunnel

The engineering press followed the construction of the Holland Tunnel with great interest. Just as work began in 1920, a long article in the *Engineering News-Record* outlined the overall plans for twin cast-iron tubes, each 29 feet in diameter. The article emphasized that all crucial design and construction questions required the approval of the outside consultants, led by Wilgus, with "the important parts of Mr. Holland's report . . . approved by a board of consulting engineers of Col. W. J. Wilgus, J. A. Bensel, Wm. H. Burr, Edward A. Byrne and J. V. Davies."[38]

Using the original estimated costs of $28.7 million, the article projected that operating revenues from tolls would be sufficient to pay off the required borrowing in just eleven years and would in twenty years produce a surplus of $66.5 million to be shared equally by New York and New Jersey. Not only an engineering marvel, the new tunnel would be a fiscal marvel as well. The tunnel was envisioned to be self-supporting; even with the increased costs, the revenue model projected not only sufficient tolls to pay for the tunnel construction but to also generate a surplus. How to properly use the surplus later became a source of bitter political infighting.

The article discussed at length the location of the tunnel, the test borings of the Hudson riverbed, and the decision to build a tunnel with two lanes of traffic in each tube rather than three. Discussion of the ventilation system quoted Holland's report: "There is absolutely no question as to the feasibility or practicability of ventilating the vehicular tunnel at a reasonable cost."[39] Obviously, the engineering audience recognized the crucial importance of the ventilation system for the world's longest vehicular tunnel. The article included a cross-sectional diagram illustrating the twin air ducts below and above the roadway.

Scientific American called the project "the world's greatest vehicular tunnel."[40] Its article discussed a problem encountered in building the Pennsylvania Railroad tunnel, which involved the completed tunnel's alarming tendency to rise and fall with the tide even though the tunnel rested below the Hudson riverbed. Engineers figured out that the cast iron tunnel lacked sufficient weight to be unaffected by the tide. By comparison, the Holland Tunnel specifications included a 19-inch-thick concrete lining, resulting in a weight of $8\frac{1}{2}$ tons per running foot, ensuring that the Holland Tunnel would avoid the tidal problem encountered by the Pennsylvania Railroad tunnel.

By May 1922, with all the final design issues decided upon and contract specifications approved, the work of the Board of Consulting Engineers drew to a close. Responsibility for the completion of the construction rested with chief engineer Clifford Holland and his staff. George Day, chairman of the New York State Bridge and Tunnel Commission, wrote to Wilgus to inform the board that their services would end on May 31. He added his praise for their work: "The work of design, involving as it did, questions of great engineering difficulty, was completed with great credit to the members of the Board."[41] Day privately asked

Wilgus to continue to serve as a consultant to Holland on a per diem basis at $150 a day. Wilgus politely declined.

Completion of the Tunnel

As the completion of the tunnel approached in 1927, both New York and New Jersey prepared to celebrate another monumental engineering triumph. Tragically, Clifford Holland did not live to see the completion of the tunnel soon to be named in his honor. In October 1924, he died of a heart attack. Friends and fellow engineers attributed his sudden death to overwork and stress connected with the massive tunnel project. Wilgus attended Holland's funeral and stayed in touch with Ole Singstad, who followed Holland as chief engineer and saw the project through to its conclusion.

To test the tunnel's newly completed ventilation system, the engineers set off a "gas bomb" in one of the tubes, then measured how long the system took to remove the gas. This test, which Ole Singstad conducted on March 15, used just a small number of the eighty-four exhaust fans. To dramatize the event, Singstad invited a group of officials and journalists into the tunnel to observe the test. They all marveled at how quickly the fans removed the smoke. One reporter wrote that the "flashlight powder passed through the roof vent so quickly that it left no odor in the tunnel. . . . Engineers set off two three-minute smoke bombs. . . . The air cleared completely within a minute."[42] Singstad pointed out that the ventilation system normally operated with twenty-eight blowers in reserve. For the world's longest vehicular tunnel, the ventilation system included significant reserve capacity to meet any emergency.

As always, once the tunnel was finished the law of unintended consequences loomed. In all of the planning for the tunnel and the experimental work carried out by the Bureau of Mines to determine the amount of carbon monoxide produced by vehicle engines, the bureau did not include buses in the testing program. Now as the project became a reality, local bus companies petitioned to operate buses through the new tunnel. With a tunnel under the Hudson, a whole new component of the port's transportation system suddenly became feasible. In January 1927, the Nevin Bus Line of Manhattan petitioned Jersey City and the New York Public Utility Commission to operate bus service between Jersey City and Manhattan through the new tunnel.[43] Other bus companies followed the lead of Nevin, and soon the railroads faced another major competitor for commuter traffic within the port of New York. Of course, no one foresaw the congestion that the addition of hundreds of buses to the streets of Manhattan would cause, once the industry expanded. Eventually, the Port Authority constructed a mammoth bus terminal to consolidate all service on the west side of Manhattan.

In addition to the additional traffic load that bus service in the tunnel represented, the New Jersey State Highway Commission now rushed to complete a number of highway projects to add new highway connections to the Holland

Tunnel entrance in Jersey City. Wilgus had envisioned the tunnel simply connecting Canal Street in lower Manhattan with the local streets across the Hudson. But trucking and bus interests demanded much better access to the tunnel than could be handled by Jersey City's local streets—something neither Wilgus nor the Bridge and Tunnel Commission had anticipated.

In 1927 the New Jersey State Highway Commission reported work under way totaling over $9 million, including a major viaduct in Jersey City from the heights down to the tunnel entrance on the waterfront.[44] The Highway Commission planned to create an "interstate route" to the Holland Tunnel. After the expenditure of more than $41 million to build the tunnel, highway engineers pushed to expand access to the tunnel beyond the narrow confines of the local city streets in Jersey City. As the state of New Jersey continued to build a statewide highway system, it seemed logical, from the perspective of the highway engineers, to connect the expanding system to the tunnel under the Hudson. Soon the Holland Tunnel formed a crucial link in the ever-growing highway system within the port of New York and the metropolitan region as the auto and truck age gathered momentum.

Another unintended consequence of the opening of the tunnel involved the railroads. Speaking before the Broadway Association just before the official opening, the tunnel's chief engineer, Ole Singstad, warned that the railroads planned to abandon their expensive lighterage system and move their freight by truck through the Holland Tunnel. Singstad said that he feared trucks' jamming the tunnel. Although Wilgus and the consulting engineers had estimated a shift of some traffic to the tunnel from the Hudson River ferries, their projections included only 20 percent of the freight that was being carried by lighters and barges. Singstad described the plans of the railroads as an "uncalculated burden which the railroads will throw upon the Holland Tunnel . . . [which] may tax the capacity of the twin tubes to the utmost."[45] The railroads planned to require the Manhattan-based hauling companies to send their trucks through the tunnel to pick up freight at the New Jersey train yards rather than use the Hudson River piers. Ever uncooperative, the railroads viewed the opening of the new tunnel as an opportunity to save money even at the cost of creating a traffic nightmare, with thousands of trucks each hour crowding out all other traffic in the new tunnel.

The Tunnel Opening

On Sunday October 9, 1927, the *New York Times* devoted a special section to the opening of the tunnel, describing it as a "modern marvel" and predicting that 15 million automobiles a year would travel through it. The *Times* placed the tunnel, which it hailed as a triumph of "engineering science," in a long line of engineering achievements and pointed out, "On Nov. 13 Manhattan Island is to be connected with the United States."[46] A sobering note reported the final cost of the project as $48 million, far above the first estimates prepared by the Board of Consulting Engineers. If traffic and toll volume fell behind estimates, the two states would be forced to use general revenues to pay off the bonds sold to finance construction.

The Holland Tunnel officially opened at 12:01 p.m. on November 13, 1927. The governors of New York and New Jersey officiated at the opening, which was attended by dignitaries and invited guests. In the first seven days of operation, 158,062 vehicles passed through the tunnel, generating $82,421.70 in revenue.[47] From the very first day of operation, the Holland Tunnel proved to be a cash machine. An insatiable demand for a route across the Hudson ensured its financial success. In the first two weeks, 334,598 vehicles used the tunnel and 602,339 in the first month. At a rate of over 600,000 vehicles a month, traffic for the first year exceed the five million vehicles used in the planning models to cover costs. If early traffic levels continued, the Holland Tunnel revenue would exceed estimates by a substantial margin, creating a large surplus beyond funds needed to cover all costs, both financing and ongoing maintenance.

On the first anniversary of the opening of the tunnel, the *Times* reported the first year of operations to be a complete success. The financing model had been based on 5 million vehicles a year; the first year saw 8.5 million vehicles use the tunnel. Each day approximately 5,000 trucks and 600 buses traveled through the tunnel. Lighterage traffic declined significantly as more and more freight passed under the Hudson rather than across the river. With more commercial traffic, the *Times* asked the rhetorical question: "With the Holland Tunnel gradually given over to commercial traffic, where will the passenger cars go?"[48] The answer: build more tunnels and, if needed, a bridge! No one mentioned returning to Wilgus's plan for a freight tunnel under the Hudson River. With the great success of the Holland Tunnel, the future seemed to promise that all traffic prob-

Holland Tunnel opening ceremonies, November 1927. In the middle of the tunnel, Governor Al Smith of New York, left, shakes hands with Governor Arthur Moore of New Jersey.

lems in the port could now be solved by more tunnels, more bridges, and more highways.

The Port Authority Gains Control of the Holland Tunnel

When the Port Authority was first established in 1921, its explicit purpose was to improve the distribution of freight by rail in the port of New York. The enabling legislation passed by the states of New York and New Jersey did not direct the new agency to build tunnels and bridges for vehicular traffic. In 1921, the Port Authority published its *Comprehensive Plan for the Development of the Port District*, a plan that focused exclusively on improving rail freight service throughout the New York metropolitan region. As Jameson Doig points out, "The need to solve the region's freight problem had been the primary motivation for creating the Port Authority, and it was widely understood that this was a *railroad* problem, requiring a railroad solution."[49] The 1921 plan called for freight tunnels under the Hudson connecting a large rail yard in New Jersey directly to a series of new freight terminal buildings in lower Manhattan modeled after Wilgus's plan proposed a decade earlier. The plan also included another freight tunnel to be built under the Upper Bay in New York harbor, connecting freight yards in Bayonne, New Jersey, with the Long Island Railroad facilities in Bay Ridge in Brooklyn.

While the Port Authority's comprehensive plan presented a rational plan to improve the chaotic and expensive rail freight system in the harbor, success rested on the cooperation of the railroads. The Port Authority was not empowered to raise money directly, depending instead upon annual appropriations from New York and New Jersey. As early as 1924, the governors of New York and New Jersey proposed that the authority take over the Holland Tunnel. They argued that one part of the mission of the Port Authority was "to simplify rail to warehouse transfer."[50] Officials of the Port Authority supported the idea: they believed the tunnel would generate a significant surplus which could then be used to fund additional transportation projects, giving the authority financial independence.

Wilgus immediately condemned the idea. He called the proposed transfer a "grave blunder," especially since the tunnel was only two-thirds completed. In an address to the American Society of Civil Engineers, he pointed out that two states had financed the project and anticipated the amortization of the costs within twenty years, and then "the use of the tunnel may be made free."[51] What a quaint idea—once the tolls paid off the construction bonds, the public would be entitled to travel through the tunnel without charge.

Nevertheless, with the opening of the Holland Tunnel in 1927 and its immediate success, discussions continued about either having the Port Authority take over the tunnel or creating a new Interstate Tunnel Commission to operate the tunnel. New York's governor, Franklin D. Roosevelt, supported the interstate commission plan, as did Governor Larson of New Jersey and the New Jersey legislature. Many others feared that the Port Authority, with access to the surplus revenue of the Holland Tunnel, would embark on a series of other projects, using the surplus as a "revolving fund" to finance further borrowing. Despite the opposition of many local officials, especially in New York City, in April 1930, both states passed legislation turning over the Holland Tunnel to the Port Authority.[52]

Almost immediately, the Port Authority announced plans to build a second vehicular tunnel under the Hudson without the need for any public funding. The authority planned to use the surplus toll revenue from the Holland Tunnel to finance construction of the second tunnel. From this moment forward, the Port Authority gained increased political independence and became self-financing. Future projects of the authority included the Lincoln Tunnel, the George Washington Bridge, the region's three major airports, and eventually the World Trade Center. Once it captured the surplus revenue from the Holland Tunnel for its own use, the Port Authority abandoned its original mission to improve rail freight transportation in the port of New York. From then on the agency focused on providing improved access for the truck, the automobile, and eventually the airplane. For the railroads, the new mission of the Port Authority ensured the slow death of the port's rail system.

Wilgus and the Holland Tunnel

Wilgus's opposition to the Port Authority's taking over the Holland Tunnel would not be his last clash with that agency. His original vision of the tunnel as a link between Canal Street in Manhattan and the local streets in Jersey City faded as the traffic exceeded even the most optimistic projections. Despite the overwhelming success of the vehicular tunnel, including its use in transporting freight by truck, he continued to argue for his freight tunnel under the Hudson as the only logical and truly efficient way to move freight within the port. Even with the Holland Tunnel, New York's transportation problems remained. The tunnel merely transferred the chaos and congestion on the waterways and the piers along the shoreline to the city streets. Trucks using the Holland Tunnel created the never-ending gridlock on Manhattan's streets that continues to this very day.

Despite concerns about the impact of the Holland Tunnel on the traffic congestion on the city streets, Wilgus took great pride in the role he played in the project. "It was a great piece of work on which we were engaged—the first to unite New York City and the mainland by land means of vehicular traffic free from interruption by water carriage," he wrote in his memoir.[53] At the time the Holland Tunnel was completed, all of the city's bridges and other tunnels—the East River bridges and the subway tunnels—united Manhattan and the other New York City boroughs. Only the Holland Tunnel connected New York directly to the "mainland" across the Hudson River in New Jersey.

When the New York Central Railroad opened Grand Central Terminal in February 1913, Wilgus's name did not appear in the official program. Whether he even attended the ceremony remains unknown. Over the years, he continued to harbor bitterness at not being publicly recognized when his ideas had set the entire project in motion.

With the Holland Tunnel came invitations to attend all the opening celebrations along with official letters lauding his contributions. In 1924, he received an invitation to attend the "holing through" ceremony for the first tube after workers had tunneled from both shores simultaneously. Very careful measurements had ensured that the two tunnel bores met precisely, and this called for a celebration. A year later, Ole Singstad sent Wilgus a souvenir tray from the ceremony, which he kept prominently displayed in the study at his home in Vermont.

As the official opening of the tunnel approached, Wilgus received a number of other invitations, including one for the gala dinner to celebrate completion of the tunnel. The invitation arrived in Vermont, and he sent a telegram with regrets. Flooding in Vermont prevented his attendance. In his papers, he kept the official invitation to the opening ceremony, which included a pass to "admit bearer to special stands at entrance plaza" and a badge with the logo "Official Party" dated November 12, 1927.

When the Board of Consulting Engineers completed its work, Wilgus received an official letter from the New York State Bridge and Tunnel Commission expressing "deep appreciation of the splendid service rendered by him in preparation of the design and plans for the tunnel."[54] As usual, Wilgus added this letter to his extensive collection of citations for his engineering work. Ole Singstad sent a personal letter to Wilgus also praising his contribution: "You may well take personal pride in the part which you have played in the great project . . . during the trying times when the plans were formulated and when it was necessary to have a firm belief in the soundness of the plan and the courage to fight that conviction."[55]

A final tribute touched Wilgus deeply. In August 1930, he sent a letter of thanks to Ole Singstad when the Bridge and Tunnel Commission included his name on a bronze plaque affixed to the Holland Tunnel's entrance on Canal Street, noting with pride, "It is a matter of gratification to me that my name has been permitted to appear as one of the factors in such a noteworthy project."[56] At Grand Central Terminal, to this day no plaque includes the name of William J. Wilgus.

7

JOINING STATEN ISLAND TO NEW YORK CITY

The Narrows Tunnel

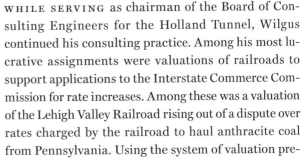

WHILE SERVING as chairman of the Board of Consulting Engineers for the Holland Tunnel, Wilgus continued his consulting practice. Among his most lucrative assignments were valuations of railroads to support applications to the Interstate Commerce Commission for rate increases. Among these was a valuation of the Lehigh Valley Railroad rising out of a dispute over rates charged by the railroad to haul anthracite coal from Pennsylvania. Using the system of valuation prescribed by the ICC, Wilgus directed over 100 assistants in an inventory of the railroad's property. He calculated the costs of reproducing the railroad at $324,478,300.[1] After his testimony before the ICC on behalf of the Lehigh, the railroad's general counsel argued that, based on Wilgus's testimony, the railroad was undercapitalized and its current rates did not provide a fair return.

Throughout the 1920s, Wilgus conducted evaluations for a number of other railroads, including the Bangor & Aroostook; the Toledo, St. Louis and Western; the Chicago & Alton; and the New York, New Haven, & Hartford. With the earnings from these consulting assignments Wilgus purchased property along the Connecticut River in Ascutney, Vermont, and built a country retreat that he called "Iridge."

In New York tension continued between the city and the newly established Port of New York Authority. City politicians, led by Mayor John Hylan, opposed the Port Authority on the grounds that the unelected, quasi-public agency usurped local control and threatened the power of elected officials to direct transportation projects in the city and the port. With the publication of a first series of plans, the authority had proposed an underground freight subway for lower Manhattan connected to the railroads in New Jersey by tunnels under the Hudson River, identical in many ways to Wilgus's small-car freight subway. In addition to the freight subway, the authority had also proposed an "inner" belt line railroad in New Jersey from Piermont, above the Palisades, to Bayonne and Perth Amboy, opposite Staten Island. To move freight more efficiently across the harbor, the authority also planned to build a freight tunnel under the Upper Bay linking Greenville in New Jersey to Bay Ridge in Brooklyn, currently served by the car floats of the Pennsylvania Railroad.[2]

While the New York politicians could not prevent the railroads from participating in a plan for an inner-belt railroad in New Jersey, they viewed the plan for a freight tunnel under the Upper Bay as a direct threat to the inherent power of the city government. In addition, neither the Port Authority's belt line plan nor the Upper Bay tunnel provided transportation improvements for Staten Island. Mayor Hylan and the New York City political leaders felt a strong obligation to include Staten Island in the city's efforts to improve mass transit and freight distribution. Hylan, with the support of Staten Island politicians and business leaders, proposed that the city of New York build a tunnel under the Narrows separating Staten Island from Brooklyn. Planned to accommodate both passengers and freight trains, the "Narrows Tunnel" would end Staten Island's isolation from the other boroughs.

The Plan for a Narrows Tunnel

In the summer of 1921, Mayor Hylan and the city's representatives in Albany introduced legislation requiring New York City to construct a freight and passenger tunnel under the Narrows between the boroughs of Richmond (Staten Island) and Brooklyn.[3] Further, the law required construction of the tunnel to begin within two years. The New York City Board of Estimate immediately appropriated money to begin planning and directed the board's chief engineer, Arthur Tuttle, to recruit a group of eminent transportation engineers to serve as consultants. Hylan believed the legislation ensured construction of the Narrows tunnel and dramatically reasserted the city of New York's local control over transportation improvements.

Tuttle turned to William Wilgus to serve as a consulting engineer for the tunnel project. In part, this was because Wilgus had, more than a decade earlier, in 1910, proposed a plan for a belt line railroad in the port of New York, including a rail tunnel connecting Staten Island to Brooklyn.[4] Wilgus described Tuttle's invitation as a new assignment that "unexpectedly came my way, to yield me compensation . . . and embroil me in a political fight."[5] As with all of his professional endeavors, he set to work to prepare a detailed report supporting the tunnel plan and began to publicly advocate for the Narrows Tunnel.

Wilgus realized that the Narrows Tunnel consulting job allowed him the opportunity to revisit his 1910 plan for an outer belt line railroad. He immediately suggested expanding the scope of the Narrows Tunnel project to include a rail line across Staten Island and a bridge over the Arthur Kill. Adding a railroad bridge over the Arthur Kill would create a direct rail link from New Jersey, across Staten Island, and under the Narrows to Brooklyn. The Narrows Tunnel offered the opportunity to connect the railroads in New Jersey directly to the Long Island Railroad in Brooklyn by crossing Staten Island. From Brooklyn, freight trains could ride on the New York Connecting Railroad through Queens and over the Hell Gate Bridge to the Bronx. The Narrows Tunnel provided the key to creating the eastern portion of the belt line railroad he had first proposed a decade earlier. Freight destined for Staten Island, Brooklyn, Long Island,

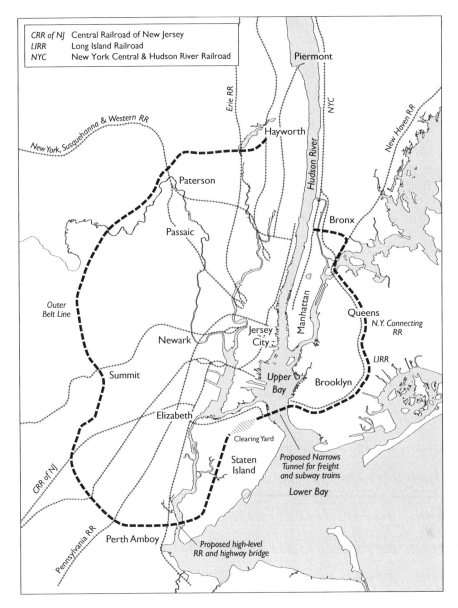

CRR of NJ Central Railroad of New Jersey
LIRR   Long Island Railroad
NYC   New York Central & Hudson River Railroad

Wilgus's 1921
plan for the
Narrows freight
and passenger
tunnel with
connections to
outer belt line
in New Jersey.

Queens, the Bronx, and New England no longer would need to be floated across the harbor, a costly and inefficient way to move freight.

To build support for the city's tunnel project, Wilgus spoke before the New York chapter of the American Society of Civil Engineers in December 1921. He explained that the tunnel represented more than just a "local improvement," as the Port Authority recently charged. Both the time and cost of delivering freight across the harbor would be reduced, Wilgus argued, "by the diversion of freight around instead of delivery in Manhattan, followed by trucking to the outlying boroughs."[6]

Wilgus followed up his talk before his fellow engineers with an article in the prestigious *Scientific American*. Here, he outlined the problems created for the entire country by the cumbersome and costly system of freight handling in the port. For two centuries the port and the Hudson and East Rivers had provided New York with unique advantages, especially in the era of sail, when freight arrived by ship or canal barge. With the rise of the railroads, these advantages became liabilities; freight arriving by rail on the New Jersey side of the harbor had to be hauled across the water to Manhattan or Brooklyn. He enumerated the threats to the continued preeminence of the port of New York:

1. The apathy of its citizens who, like the fabled hare, have not realized in greatness lurks the seed of its own decay—an overweening confidence in the maintenance of supremacy without effort.
2. The presence of a physical barrier of the first magnitude between rail terminals on the New Jersey mainland and the island city of New York; and
3. The existence of the boundary between the states which politically separates the two parts of the common port.[7]

In his analysis the Narrows Tunnel formed one part, albeit a crucial part, of the larger plan for a belt railway to encircle the entire port of New York. With the addition of a Narrows Tunnel and the belt line, quick and efficient freight transfers between all of the railroads would be possible, significantly lowering costs and ensuring the port's continued dominance.

Wilgus recognized a serious challenge: the cost of the Narrows Tunnel and belt line railroad could not be justified unless "revenue from prospective users should far exceed anything that could be expected from the very meager interborough passenger service."[8] Testifying before the Board of Estimate in January 1922, Wilgus estimated the cost of the tunnel at $110 million, including the tracks and rail yard on Staten Island and a viaduct over the Arthur Kill connecting to the rail network in New Jersey.[9]

Once again, plans for an innovative solution to the port's transportation problems rested on the cooperation of the railroads. Revenue depended on the railroads' abandoning their expensive lighterage and car float systems and diverting freight traffic by rail through the Narrows Tunnel. If the railroads participated, projected freight fees would be more than sufficient to cover the bonds necessary to finance construction. Once again, the financial model depended upon a rational, cost-benefit analysis on the part of the private railroads. As always, the railroads proved incapable of such thoughtful and far-sighted behavior.

Reaction to the Narrows Plan: "All Hell Broke Loose"

Wilgus typically described all of his plans and proposals using the dispassionate language of the engineer; his reports provided models of careful, deliberative planning with conservative cost estimates. Political in-fighting and corporate intrigue remained in the background. With his engagement as a consultant to the Narrows Tunnel project, he became embroiled in a giant battle over the future of the port and rail freight distribution. Of course, in a long and successful career he had faced a number of controversies, including the bitter internal fight at the New York Central over responsibility for the Woodlawn Wreck that ended his career with the Central. With the Narrows Tunnel, a public battle unfolded, and as he succinctly summarized the conflict, "All hell broke loose." Hell came from three directions—from the Pennsylvania Railroad, from the Port Authority, and from New York's former governor Alfred E. Smith.

The Pennsylvania Railroad viewed the Narrows Tunnel as a threat to its "natural territory." Railroads with established service to a particular geographical area opposed any plans to build competing facilities and offer competitive freight service. Once the Pennsylvania Railroad gained control over the Long Island Railroad in 1900, it regarded Brooklyn, Queens, and all of Long Island as its domain.

The Pennsylvania's freight service to Brooklyn and Long Island involved an expensive and convoluted system of lighterage and car floats to haul freight across the harbor. At Greenville, New Jersey, the railroad had built a rail yard and piers for car floats. Freight cars destined for Brooklyn and Long Island were loaded on car floats and then towed across the Upper Bay to Bay Ridge in Brooklyn. Unloaded there, the freight cars were then taken via the Long Island Railroad to destinations in Brooklyn or Long Island. Since the Pennsylvania could not charge New York customers a higher rate, the delivery of freight across the harbor cost the railroad a small fortune.

Wilgus, logical as always, reasoned that the Pennsylvania would realize significant cost savings by using the new tracks across Staten Island and the Narrows Tunnel. Even after paying the fees to use the new rail connection, the railroad would realize a savings over the current cost of freight service across the harbor.

As soon as Wilgus agreed to serve as a consultant to the tunnel project, he wrote to all of the railroads and described the cost savings to be gained with the construction of the belt railway and the Narrows Tunnel. He received an immediate reply from the assistant chief engineer of the Pennsylvania Railroad, R. B. Temple, who laid out a series of objections to both the belt line and the tunnel. Temple's objections focused on current estimates of the costs of hauling freight and the projected savings from using the tunnel. In addition, he argued that Wilgus overestimated the amount of freight to be "diverted" over the new rail link to Brooklyn. Railroads with facilities in the "northern" section of the New Jersey side of the harbor gained little and would continue to use lighters and car floats. Only the "southern" group of railroads would achieve modest cost savings using the new rail link; therefore, only a modest amount of freight would be diverted to the new system. For these reasons, Wilgus's figures for the amount of freight to be carried overestimated future traffic and revenue. The tunnel project and belt line would never be self-sustaining, and the City of New York would ultimately be forced to cover the cost of financing the project and then pay operating costs.

Wilgus responded to the Pennsylvania's objections with a detailed nine-page letter. He challenged the cost figures Tuttle used for freight delivered to Brooklyn and argued that the other railroads incurred even higher costs than the Pennsylvania's of $6.56 per freight car. In addition, he objected to the Pennsylvania's estimates for switching costs in the new Staten Island rail yard. Wilgus concluded that for the "southern" railroads, "the project from the start would be at least self-supporting" and would have "future potentialities of great value."[10] Future growth always figured into Wilgus's cost models. He believed that great transportation projects should be judged both on current savings and future benefits. The New York Central Railroad eventually realized enormous profits from developing the air rights over the underground train yard to the north of the new Grand Central Terminal. Those profits accrued decades after the completion of the project, however, requiring imagination and foresight on the part of railroad executives.

With the Narrows Tunnel, Wilgus firmly believed that expanded freight traffic from New Jersey across Staten Island and through the tunnel to Brooklyn would be inevitable. Both the railroads and the city of New York stood to gain. Wilgus concluded with a plea to the railroad: "There are so many other interests to be considered than those of the railroads." He then enumerated the benefits for all: "The City of New York, embracing some six million people, feels that it is entitled to an all-rail connection with the mainland for its commercial development and the welfare of its citizens." He added an argument for national defense in time of war: "The nation needs a means of coordinating the railroads . . . so that in time of war our armies and their supplies may be drawn from any direction . . . for national defense." Even an appeal to national defense fell on deaf ears with the Pennsylvania and the other railroads. Once again, self- interest prevailed over reason. For the third time in his career, Wilgus's efforts to gain cooperation from the railroads to improve freight delivery in the port and city of New York failed.

Opposition to the Narrows Tunnel also came from the newly formed Port Authority, which planned to build a freight tunnel under the Upper Bay connecting Greenville, Jersey City, with Bay Ridge, Brooklyn. Julius Cohen, general counsel and chief advocate for the authority, immediately started a battle to sway public opinion. At the authority's new offices at 11 Broadway, he staged an elaborate exhibit of the plans for the Upper Bay tunnel and invited the press to a series of news conferences. Alfred Smith, former governor and a long-time political power in New York City, attended one of the conferences to add gravitas in his new role as one of the three New York commissioners of the Port Authority.

Cohen's exhibition emphasized the advantages of the Port Authority's tunnel. He snidely remarked that of the numerous municipalities in the authority's regional jurisdiction, only the city of New York opposed the plans. It seemed that New York's duly elected mayor and Board of Estimate opposed the authority's grandiose plans based on the fear "that the autonomy of the municipality will be sacrificed."[11] From Cohen's perspective, the city of New York, even if it was the largest and most prosperous city in the country, was just another of the numerous municipalities with which the authority had to deal.

The Port Authority's exhibit used a cost figure of $11.34 per freight car for its

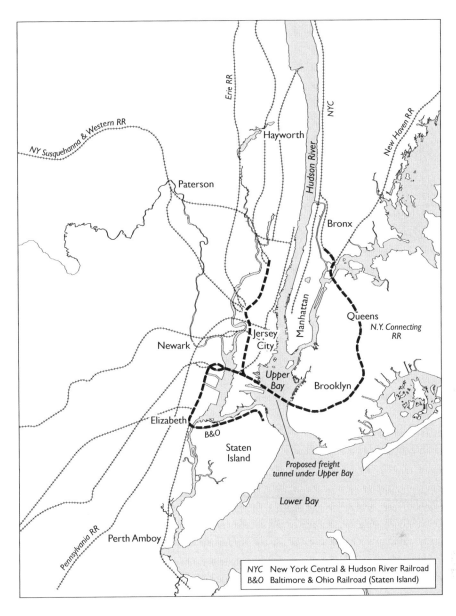

The Port Authority's 1921 proposed freight tunnel under Upper Bay from Greenville, New Jersey, to Bay Ridge, Brooklyn.

NYC	New York Central & Hudson River Railroad
B&O	Baltimore & Ohio Railroad (Staten Island)

planned freight tunnel versus $15.17 to deliver a freight car to Brooklyn over the proposed Staten Island belt line and Narrows Tunnel. Cohen argued that the authority's cost of $11.34 resulted from a careful analysis, but in fact, he had lifted the cost directly from the chief engineer of the Pennsylvania Railroad, who used the same figure in his earlier letter to Wilgus. Cohen simply needed to cite a lower cost to discredit Wilgus and the New York City plan.

Mayor Hylan and New York's other elected officers did not stand idly by as the Port Authority stepped up criticism of the Narrows Tunnel plan. John Rooney, ex-president of the New York Court of Claims and an important city politician, wrote a long article for the *New York Times* defending the city's plan. Rooney raised the issue of "local control" as a prime reason for the city's opposition:

"Such a proposition [the Port Authority tunnel] must arouse the opposition of the elected representatives of the City of New York, whose duty it is to protect the city's interests and prestige."[12]

The battle over the power of the Port Authority versus the local control vested in municipal government expanded as the number of public authorities increased. Numerous authorities followed in the footsteps of the Port Authority as the state of New York and other states turned to authorities to "get things done." State leaders delegated power to authorities in the belief that authorities would exercise rational planning and bring order to the often chaotic politics of local government. As a side benefit, independent authorities with appointed rather than elected officials stood a better chance of resisting corruption than elected officials and local political machines. From rather modest authorities charged with mosquito control to behemoths like the Triborough Bridge and Tunnel Authority or the Port Authority, the delegation of power to authorities in New York and across the country came to have profound consequences. Robert Moses eventually served as chairman of a number of authorities, including the Triborough Bridge, and used the accumulated power to change New York and the region forever. For his biography of Moses, Robert Caro chose the title *The Power Broker* not because Moses won elections—he never won an election—but because of the real power he exercised. The power that emanated from these independent authorities went on for decades without the checks and balances inherent in the normal political process.[13]

Wilgus's Analysis of the Narrows Tunnel and the Belt Line

Having already thought through the advantages of a freight tunnel connecting Staten Island to Brooklyn, Wilgus drew upon his earlier studies in preparing his report to the Board of Estimate and its chief engineer, Arthur Tuttle. Late in January 1922, he submitted his preliminary report.

In a cover letter to Tuttle, Wilgus expressed his conviction that the tunnel project provided a real opportunity to dramatically improve freight delivery in the port of New York. The tunnel plan included subway trains' sharing the freight tracks under the Narrows, thus providing mass transit between Staten Island, Brooklyn, and Manhattan. He placed the Narrows Tunnel in the same category as both the Grand Central and the Pennsylvania passenger stations: "The more I consider this project the more I believe in it. Perhaps this feeling is colored by the successful outcomes of other momentous enterprises in which I have taken a part . . . or many other advances in the art of transportation that have been of such vast benefit to man."[14] For the "vast benefit to man"—not a modest claim indeed but one that illustrates Wilgus's faith in his own work.

Wilgus framed the project as one challenging the "status-quo." He noted that some interests, primarily the railroads, were arguing that minor improvements to the current freight delivery system would suffice, without a costly tunnel and rail project. On the other hand, a whole litany of civic groups wholeheartedly were supporting the project, including the New York Chamber of Commerce, the

Merchant's Association, the Brooklyn Chamber of Commerce, the Staten Island Civil League, and the press. He also evoked a national defense argument to support the Narrows Tunnel. To improve shipping in time of war, transportation experts had proposed dredging Jamaica Bay for a deepwater port to relieve congestion in New York harbor. These plans for Jamaica Bay absolutely depended upon a direct rail link with the mainland to be at all feasible.

Wilgus rejected a laissez-faire approach and argued for bold, innovative steps to solve the freight distribution problem. As he had so often done in his career, he rejected the small-step, piecemeal approach. The tunnel project, he told Tuttle, provided for "radical improvements in our methods of cross-harbor transportation."[15] Just as with his small-car freight subway plan of 1907, the Narrows Tunnel offered the opportunity to finally create a direct rail link between New York and the mainland across Staten Island and then to Brooklyn.

The New York State enabling legislation required the Narrows Tunnel to be "self-supporting," generating sufficient revenue to pay interest and principal on the city of New York bonds sold to pay the construction costs. Wilgus knew that local freight traffic between Staten Island and Brooklyn could never generate the necessary toll revenue to cover the significant costs of the tunnel. To be successful, the belt line and the tunnel needed customers, and those customers consisted of the "trunk lines now terminating in New Jersey." The numerous railroads with freight yards, terminals, lighters, and car floats had to be persuaded that significant savings awaited before the railroads would agree to divert their freight destined for the east side of the harbor to the belt line and the tunnel.

Arguments against the Narrows Tunnel and the belt line focused on the more roundabout path of the outer belt line in New Jersey that was proposed to connect all of the railroads. As planned, the belt lines totaled 41 miles in length, longer than the inner belt line included as part of the Port Authority's plan for a freight tunnel under the Upper Bay. Wilgus pointed out that for the southern group of railroads, which carried 80 percent of the freight tonnage floated across the harbor, the haul would be shortened by almost three miles. The distance "hauled" mattered to the railroads because freight operating costs correlated directly with distance. Shorter hauls cost less.

Wilgus's letter to Tuttle included detailed criticism of the Port Authority tunnel plan. He pointed to that tunnel's route under the widest part of the harbor, the Upper Bay from Greenville, New Jersey, to Bay Ridge in Brooklyn. He also criticized the location of the authority's planned inner belt railway, which would require the construction of a trestle across Newark Bay. A trestle across the bay would cause shipping delays, since the trestle's drawbridge would have to be opened to allow ships to pass. Wilgus enumerated eight other defects of the plan and concluded: "Their plan is more or less a fog. . . . [It] is gravely at fault in not having been based on fundamentals which are vital to the effective solving of New York City's cross-bay transportation problem."[16] To complete the criticism, Wilgus estimated costs for the Narrows Tunnel and belt line to be at least $35 million less than costs for the Port Authority's plan—a very significant sum.

To support the argument that the belt line and tunnel would generate sufficient revenue to be self-sustaining, Wilgus carefully analyzed the existing volume of freight exchanged across the harbor to Bay Ridge and the Brooklyn waterfront (table 7.1). He again divided the New Jersey railroads into two groups: a southerly group consisting of the Pennsylvania, the Baltimore & Ohio, the Central Railroad of New Jersey, and the Lehigh Valley; and a northern group that included the Erie, the Delaware & Lackawanna, the West Shore, and the Ogdensberg & Western. Coal accounted for a major share of tonnage. In the 1920s burning coal provided heat, ran the factories, powered ships, and provided the raw material for "natural" gas.

Wilgus assumed that a major share of that traffic would be diverted to the belt line and Narrows Tunnel. Charges for hauling the diverted freight cars would generate sufficient revenue to cover the construction bonds and provide for ongoing maintenance. Wilgus and the Board of Consulting Engineers had used the same model for projecting toll revenue for the Holland Tunnel. In the case of the Holland Tunnel, their estimates proved to be conservative; toll revenue exceeded projections from the day the tunnel opened. With the Narrows Tunnel, similar assumptions seemed reasonable.

For the New Jersey railroads, especially the southern group, using the tunnel between Staten Island and Brooklyn rather than floating freight across the harbor would significantly lower freight costs. The tunnel would also provide much quicker delivery of freight, another decided advantage. As always, Wilgus assumed that the railroads operated with a rational decision-making process. If a new transportation link offered lower costs and faster service, any business would seize the opportunity. The Holland Tunnel had already proved that freight shippers and motorists, when offered improved service and lower costs, rapidly switched from the ferries to the tunnel. In fact, service through the Holland Tunnel had proved so attractive that delays soon occurred and the public clam-

Table 7.1. Volume of freight traffic between New Jersey, Bay Ridge, and Brooklyn waterfront, in 1921 (tonnage)

	New Jersey railroad group		
	Southern group	Northern group	Total
Interchange via Bay Ridge	11,400,000	940,000	12,340,000
Brooklyn waterfront			
Car floats	2,370,000	1,630,000	4,000,000
Lighterage	440,000	560,000	1,000,000
Lighterage—coal	3,015,000	985,000	4,000,000
Brooklyn total	5,825,000	3,175,000	9,000,000
Grand total	17,225,000	4,115,000	21,340,000

SOURCE: "Interchange via Bay Ridge: Freight from New Jersey to Long Island Railroad Freight Yard in Bay Ridge, Brooklyn, for Distribution via Long Island Railroad," Table A, Tonnages and Cars, 1921, William J. Wilgus Papers, Manuscripts Division, New York Public Library, box 54, folder 13—Tonnages.

Table 7.2. Estimated freight costs per car using existing water routes (1922) and using proposed Staten Island belt line and Narrows Tunnel

	New Jersey railroad group		
	Southern group	Northern group	Average cost
COSTS BY PRESENT ROUTE			
Interchange via Bay Ridge	$7.03	$8.40	$7.15
Brooklyn waterfront			
Car floats	9.31	10.93	9.97
Lighterage	18.32	18.47	18.40
Lighterage—coal	14.82	14.95	14.85
Average, all destinations	$8.85	$12.19	$9.64
COSTS BY BELT LINE AND NARROWS TUNNEL			
Interchange via Bay Ridge	$3.11	$5.60	$3.33
Brooklyn waterfront			
Car floats	4.75	7.07	5.69
Lighterage	6.36	8.34	7.46
Lighterage—coal	6.39	8.73	6.96
Average, all destinations	$3.97	$7.18	$4.73

SOURCE: William J. Wilgus Papers, box 54, folder 19—Preliminary Estimates.

ored for the Port Authority to build additional vehicular tunnels under the Hudson or a bridge.

To haul all of the freight required more than 1.5 million freight cars a year, with over 800,000 interchanged between Greenville and Bay Ridge and the remainder to the Brooklyn waterfront. Wilgus estimated substantial savings for each freight car currently interchanged with either Bay Ridge or the Brooklyn waterfront (table 7.2). He calculated total savings at $7,540,800 if the new route diverted all of the freight cars to the belt line across Staten Island and then through the Narrows Tunnel.

Wilgus concluded his report by first listing the estimated costs of the project, including the belt line across Staten Island, the tunnel, and the viaduct across Arthur Kill (table 7.3). He added rail tracks along the Brooklyn waterfront to enable freight cars to be delivered directly to the piers. His projected grand total was $110 million. Using 6 percent interest in New York and 7 percent in New Jersey as the cost of borrowing, he estimated yearly fixed charges at $6.9 million. Net saving for the railroads by using the Narrows Tunnel would total $9.8 million, more than needed to cover the fixed charges and in fact generating a "surplus" of almost $3 million a year.

Wilgus also noted that even if the revenue from freight fell below estimates,

Table 7.3. Wilgus's estimated construction costs
for the Narrows Tunnel project

New Jersey	
Haworth to vicinity of Metuchen (outer belt)	$25,000,000
Vicinity of Metuchen to Perth Amboy	3,500,000
Arthur Kill viaduct	3,500,000
Total	$32,000,000
New York	
Arthur Kill viaduct	$2,000,000
Arthur Kill to East End Richmond yard	8,000,000
Richmond Yard to Bay Ridge (tunnel)	43,000,000
Total	$53,000,000
Brooklyn railway	25,000,000
Grand total	$110,000,000

SOURCE: Wilgus to Arthur Tuttle, chief engineer, Board of Estimate and Apportionment, Jan. 18, 1922, 33, William J. Wilgus Papers, box 53.

the Narrows Tunnel project provided another entrance to the outer boroughs for passenger trains, a potential source of revenue. He included a reference to his foresight when planning the Grand Central project, which resulted in unforeseen benefits for the Central railroad, difficult to quantify at the time:

> If I may be pardoned for the personal allusion, the Grand Central Terminal owes its existence to a forecasting in 1902 of the brilliant possibilities that [lay] in the use of the overhead air rights then lying dormant in face of prophesies that railroad hotels never had been nor would be successful. . . . The actual results have far exceeded the favorable forecasts of the time. . . . Happily in this case the analysis has merit of reinforcing judgment. . . . With the prospects so bright for the project to be self-supporting from the start there would seem to be no question whatever as to the advisability of proceeding at once.

Wilgus finished his projections with the assertion that "the City and State of New York want it, shippers want it, and the consumer ultimately wants it."[17]

As with his proposals for the small-car freight subway and the inter-terminal belt line subway and elevated railroad, Wilgus's initial report for the Narrows Tunnel did not focus on the technical aspects of building the tunnel. Given his extensive experience with the Detroit Tunnel and the Holland Tunnel, he was confident that the technological expertise of subaqueous tunneling building had reached a point where actual construction of the tunnel did not present a major problem. Instead, the key challenges were marshalling political support and overcoming the opposition of the Port Authority and the Pennsylvania Railroad. In the case of the Narrows Tunnel, the engineering, as far as the role Wilgus played, took a back seat.

The Battle over the Narrows Tunnel

Opposition from the Port Authority continued because the city's tunnel under the Narrows threatened to make the authority's tunnel under the Upper Bay redundant. Further, the Upper Bay tunnel formed a key component of the authority's first master plan. Not moving forward with that tunnel would undermine the new agency's credibility. From the beginning, the Port Authority's critics argued the port of New York did not need a new "super" quasi-governmental entity to solve the port's transportation problems.

For the Pennsylvania Railroad both tunnels represented a threat to its dominance of freight delivery to Bay Ridge and to the Long Island Railroad, one of its subsidiary lines. In addition, the Pennsylvania also handled a major share of the freight and cars to and from the Brooklyn waterfront, where the Narrows Tunnel would connect. Of the total of 1,173,000 freight cars floated across the harbor between New Jersey and Brooklyn in 1921, the Pennsylvania handled 610,000, over 50 percent; and it also hauled more than 50 percent of freight carried by lighters.[18] The railroad considered both Brooklyn and Long Island to be its territory and did not want the other railroads to have lower costs and greater efficiency delivering freight.

Wilgus continued to press for the Narrows Tunnel. In late January 1922 he appeared before a joint committee of the New York State Legislature regarding the Port Authority's recently completed master plan. To move forward, the authority needed enabling legislation from both New York and New Jersey. Wilgus opened his testimony with a call for dispassionate, rational comparison of the two tunnel plans: "Involving as they do technical questions in which transportation engineering plays a principal part, it seems only proper that the two plans should be analyzed in such a manner that you may clearly see in what respects they differ."[19] He argued that the Narrows Tunnel's key advantage involved utilization by both freight trains and rapid transit, creating desperately needed rapid transit passenger service between Staten Island and Brooklyn. He ended his testimony with the statement: "The greatest good for the greatest number is the underlying thought of the City's proposition."[20]

Wilgus followed his first report supporting the Narrows Tunnel with a major addendum to Arthur Tuttle in February, responding to the Port Authority's criticism of the tunnel. His main argument addressed the criticism that using the belt line across Staten Island rather than the authority's tunnel under the Upper Bay added distance to the railroad's haul. Wilgus intended the report to provide the city with further arguments against the authority's comprehensive plan before the state legislature. Despite his efforts and the continued opposition of the city's representatives in Albany, the Republican-dominated legislature approved the Port Authority's master plan. While the authority now had formal approval to proceed with its plans, it lacked financing.

The battle over the Narrows Tunnel versus the Upper Bay tunnel continued through 1922, with neither side willing to compromise. On the surface, the city of New York appeared to hold the upper hand. The 1921 legislation authorizing the city to build the Narrows Tunnel included allowing the city to sell bonds for construction. The legislature's approval of the Port Authority's comprehensive plan did not include financing. The new agency did not at this time have the legal right to sell bonds, and any major transportation initiatives needed funding by appropriations from both New York State and New Jersey.

Privately, Wilgus knew that the decision about whether to construct the Narrows Tunnel hinged on politics, not on technical engineering questions or on projections of the volume of freight likely to be diverted to the new tunnel and belt line. In his memoirs, Wilgus recounted the politics of the tunnel's ultimate defeat: "The Port of New York Authority raised a hue and cry over what they judged to be an invasion of their rights and privileges. . . . Ex-Governor Alfred E. Smith . . . gave his support to the assault on the Narrows tunnel project. . . . The press of the city with few exceptions took sides with Ex-Governor Smith and the Port Authority and termed the tunnel 'Hylan's folly.' "[21]

With all this opposition it seems hard to imagine how the project could have moved forward at all. With no humility Wilgus attributed the continued efforts to build the tunnel to his own brilliance: "In breathing life into the project, by offering a way to make it a financial success, I was the one who had upset the politician's apple cart."[22]

New York mayor Hylan continued to push vigorously for the tunnel project despite the formidable public opposition of former governor Smith, now serving as a commissioner of the Port Authority, and the covert maneuvering of the Pennsylvania Railroad. Since the city of New York already had enabling legislation, it proceeded with technical planning for the tunnel throughout 1922 even as the political battles continued. In early 1923, after acquiring needed land for the right-of-way for the tunnel approaches in Staten Island and Brooklyn, the city began construction of the tunnel by sinking shafts on each side of the Narrows. On January 20, 1923, Mayor Hylan attended a ceremony on Staten Island to celebrate breaking ground for the Staten Island shaft of the Narrows Tunnel. The *Times* reported that "throngs" turned out for the occasion. Hylan took the opportunity to lash out at the tunnel opponents, calling them "corporation bandits" supported by the "lickspittle press," and praised his own efforts: "It takes a man of iron to deal with these people. . . . I am thankful I have the health to stand against that band of bandits."[23]

As the press had already pointed out the preceding year, however, the initial appropriation from the Board of Estimate to start the tunnel did not include funding for the belt line railroad across Staten Island to the Arthur Kill. The viaduct over the Kill connecting to New Jersey needed the approval and funding of both states and could not be constructed by the city alone. Without the belt line the tunnel remained just a local transportation project and would never be able to generate sufficient revenue to be self-supporting.[24]

Behind the scenes Wilgus opposed moving forward with construction of the shafts. He feared the ever-present opposition from the Port Authority and the possibility of further political problems in Albany. Hylan, Tuttle, and the Board of Estimate ignored Wilgus's pleas for caution and plowed right ahead with construction of the two shafts and with detailed design work for the tunnel itself. Always sure of his own judgments, Wilgus resigned as a consulting engineer to the Narrows project, stating, "My conscience prompted me to resign."[25] The project went forward nevertheless, and the city ultimately spent over $5 million on land acquisition and the two shafts.

More ominously, Alfred Smith ran for governor again in 1922 and won by a decisive margin. Once in office, he did not let up in his battle with Hylan over the Narrows Tunnel, nor did he forget being referred to by Hylan as among the "corporation bandits" opposed to the tunnel. Smith continued his opposition to the tunnel and supported legislation in Albany to amend the 1920 law authorizing the city to construct the Narrows Tunnel. After more than a year of debate, the state senate passed a bill in late March 1925 restricting the tunnel to rapid transit, barring freight trains from using the tunnel. Before Governor Smith signed the legislation, Hylan rushed to Albany to lobby against the legislation; he argued that "the people of the different boroughs have unmistakably expressed their preference for the city's freight and passenger tunnel."[26] His populist argument failed to persuade the Republican-dominated senate.

In a dramatic confrontation on April 21, 1925, Mayor Hylan testified in Albany in opposition to the bill banning freight from the Narrows Tunnel. Governor Smith chaired the hearing. Hylan insisted the Pennsylvania Railroad remained the real opposition to the tunnel. Julius Cohen of the Port Authority spoke in support of the legislation, as did the world-famous railroad engineer George Goethals, now a consulting engineer to the Port Authority. Smith listened to the arguments for over five hours. The press reported Smith ready to sign the legislation even if it came at the cost of "an open break between Governor Smith and Mayor Hylan."[27] Ignoring Hylan's arguments, Smith signed the law the next day.

Despite the city's expenditure of millions of dollars and after a year of construction activity already under way, the state legislature passed the legislation restricting the Narrows Tunnel to passenger trains. Without the revenue from freight service, the city knew the tunnel could never be self-supporting. Faced with the inevitable, the city announced suspension of all further work on the Narrows Tunnel. To cement his triumph over Hylan, Smith supported Jimmy Walker in the Democratic primary for mayor later in 1925. Walker defeated Hylan in the primary, effectively ending John Hylan's political career.

Ironically, in the late 1950s, with the automobile and truck age roaring forward, Robert Moses and his Triborough Bridge and Tunnel Authority announced plans, with their usual fanfare, for a suspension bridge over the Narrows. To keep costs within reason, Moses planned the bridge approaches on the land that the city of New York had acquired in the 1920s for the Narrows Tunnel. Moses also built a superhighway—Route 278—across Staten Island connecting with the Port Authority's two bridges over the Arthur Kill to New Jersey. On the Brooklyn side of the bridge this highway connected to the newly constructed Belt Parkway along the Brooklyn shore and the Brooklyn-Queens Expressway. The highway system that Moses created followed almost precisely the path laid out for the belt railway and railroad tunnel first proposed by William Wilgus in 1911.

The Port Authority and the Railroads

With the city's Narrows Tunnel brought to an abrupt halt, the Port Authority seemed to be in the dominant position to move forward with its plan for a freight tunnel under the Upper Bay from Grenville in New Jersey to Bay Ridge in Brooklyn. With the support of Mayor Walker and Governor Smith, the authority needed only the cooperation of the private railroads to proceed with the tunnel. Political leaders in both Albany and Trenton promised to secure the necessary financing once the agency signed agreements from the railroads to divert their freight to the new tunnel. The Port Authority's annual report for 1926 noted that the tunnel's success depended upon each railroad's "sacrifice of individual advantage" in exchange for gains in efficiency for the entire port.[28] Once again, a major improvement in the freight transportation system for the port of New York depended upon the cooperation of the railroads. Once again, the railroads refused to cooperate.

The Port Authority now was to learn the same painful lessons William Wilgus had encountered with his proposals for the small-car freight subway and inter-terminal belt line more than a decade earlier: no railroad serving the port voluntarily surrendered the smallest of advantages. For decades the New York Central fought the city of New York tooth and nail to defend its track rights on the city streets on the west side of Manhattan. As long as the railroad continued to make money hauling freight to lower Manhattan, the "public be dammed," to borrow a phrase from the Commodore's son, William Henry Vanderbilt.

In the end, none of the rail freight improvements planned as part of the authority's first comprehensive plan, including the freight tunnel under the Upper Bay, ever moved forward. As Jameson Doig, the eminent historian of the Port Authority, summarized: "Most of the Port Authority's rail-freight hopes and regional designs crashed to earth, the victims of the reluctance of the private rail corporations to change their habitual ways." The intransigence of the railroads proved to be the downfall of the agency's tunnel plan—but also, ultimately, the downfall of the railroads' monopoly as well. In the 1920s trucks increasingly gained a greater and greater share of freight delivery in the port. Rebuffed by the railroads in its founding mission to improve rail freight transportation in the port of New York, the authority turned its energy to building tunnels, bridges, and airports. "By the 1930s," Doig continued, "the Port agency's interests were no longer with that aged technology. Her leaders had at last embraced the benefits of the motorized truck and automobile."[29] Of course, the "aged technology" referred to the railroads.

Eventually, truck transportation failed to increase efficiency and led to ever-increasing congestion on the highway system in the port district. A 2009 study by the New York Metropolitan Transportation Council found that 35,254 freight trucks cross both the George Washington Bridge and the Verrazano Narrows Bridge every day, a nightmare indeed. An additional 8,061 trucks, on average, travel through the Lincoln Tunnel each day.[30] Today, the railroads distribute just 3 percent of all freight in the New York region. In the 1920s when the battle over the Narrows and Upper Bay rail tunnels flared, no one could have imagined the congestion and inefficiency created by abandoning rail freight delivery in favor of the truck.

In the early 1990s, Congressman Jerrold Nadler, representing Manhattan and Brooklyn, introduced legislation to fund a study to revive the plan for the rail tunnel from Greenville to Bay Ridge under the Upper Bay. With money from Congress and the city of New York, the Cross Harbor Freight Movement Major Investment Project investigated the cost of finally creating a direct rail link between Brooklyn and the mainland.[31] Environmental concerns arose, as did cost issues; and at present, a decision to proceed with the tunnel, estimated to cost billions of dollars, remains uncertain. On the other hand, in the new century no one argues for building more bridges and highways as a solution for New York's transportation woes. Once again, the railroad offers a way to increase efficiency, lower costs, and reduce congestion in the delivery of freight in the New York metropolitan region. Almost a century ago, William Wilgus warned that the future prosperity of the region depended upon improving the delivery of both passengers and freight. Time and again he presented innovative rail solutions for overcoming the inherent challenges of the port's unique geography. With the rise of the Port Authority and the planning decisions of Robert Moses, the automobile and highway age triumphed. In the beginning of the twenty-first century, the soundness of Wilgus's plans to use the inherent efficiency of the railroad may again find resonance.

The Regional Plan Association and the George Washington Bridge

In the late 1920s, Wilgus received an invitation to serve as a consultant to New York's regional planning organization to advise the agency on transportation matters. The invitation came from the chairman, Colonel Frederic A. Delano, who had served with Wilgus in France during World War I. Wilgus and Delano remained friends after the war; Delano admired Wilgus's plans to improve transportation in the New York metropolitan area. In 1925, the planning organization, officially named the Committee on the Regional Plan of New York and Its Environs, developed the first comprehensive plan for both New York City and its surrounding metropolitan region. In 1929 the Russell Sage Foundation funded a permanent organization, the Regional Plan Association of New York, to carry on the planning work of the committee.

Beginning with the completion of Central Park, designed by Frederick Law Olmsted and Calvert Vaux, New Yorkers came to realize that the city's rapid, chaotic, and totally unplanned growth resulted in both grandeur and squalor. The new professional "city planners" such as Robert Pope and Charles Mumford Robinson called for the development of master plans to bring some order out of the chaos in cities like New York. Inspired by the rebuilding of Paris by Baron

Georges-Eugène Haussmann and by the White City at the World's Columbian Exposition in Chicago in 1893, a "City Beautiful" movement began, calling for systematic planning to solve the problems of the American city. That planning included improvements to mass transportation. The first master plan for New York was prepared in 1907 by the New York Public Improvement Committee. The Regional Plan Association followed in the tradition of the City Beautiful movement but expanded the vision of the Improvement Committee to encompass the entire metropolitan region as well as the core city.

Frederic Delano, chairman of the Regional Plan Association, asked Wilgus for advice on the association's response to the Port Authority's efforts to build a massive highway bridge over the Hudson River at 178th Street. With the immediate success of the Holland Tunnel, the Port Authority began planning for the long-awaited bridge across the Hudson. George Goethals, a long-time advocate for a Hudson River bridge who was serving as a consultant to the Port Authority, urged the authority to move forward and build a bridge. Goethals preferred a bridge further south, as had the world-renowned engineer Gustav Lindenthal, with his plans for a massive bridge to be located between 14th and 23rd Streets. Bowing to strong opposition from midtown business interests, the authority decided to build a bridge to connect Fort Lee in New Jersey with upper Manhattan miles to the north of lower Manhattan, the center of New York's business and industry.

In 1923, O. H. Amman, another brilliant engineer in the age of the engineer, submitted preliminary plans to the Port Authority for a bridge across the Hudson that would be the longest suspension bridge in the world. With the support of the governors of New York and New Jersey, the authority worked to gain passage of the necessary enabling legislation in 1925 and 1926. In July 1925 Amman became the chief bridge engineer for the Port Authority and proceeded to complete the complex engineering and final design for the bridge at 178th Street.[32]

The Regional Plan Association supported plans for the bridge. In a letter to Wilgus in February 1926, Thomas Adams, the association's director of plans and surveys, discussed alternative locations for the bridge between 192nd and 178th Streets but emphasized the importance the association placed upon the bridge: "Our attitude is that we must accept [the Port Authority's] final judgment" on the location of the bridge "because the great thing is to get the bridge." While the Regional Plan Association wanted the bridge, Wilgus wanted the bridge design to include provisions for accommodating railroad tracks as well as vehicular traffic. Adams anticipated resistance on this point from the Port Authority, explaining to Wilgus that the authority would "raise the objection that in the absence of any definite plan for carrying out the railroad and transit lines to form the belt line system they should not go to the expense of building a bridge to meet conditions that might never arise."[33]

As part of the process of gaining approval for the bridge, the Port Authority conducted a series of public hearings and meetings with various interested parties throughout 1926. The authority met a number of times with the Regional Plan Association, and prior to each meeting the association sought Wilgus's counsel. Wilgus did not give up on his idea for railroad tracks on the bridge, despite Adams's earlier caution on the subject. In March Wilgus wrote to Adams and discussed the costs of providing for railroad tracks, emphasizing that "the cost of providing for the additions to the structure necessary for carrying of railroad trains is not as important as at first blush. . . . Sight certainly should not be lost of the railroad needs at the proposed crossing."[34] In a follow-up letter Wilgus cautioned Adams that the railroads would in all likelihood not support including provisions for railroad tracks on the bridge.

In a letter in early September 1926, Wilgus again stressed to Adams the absolute importance of adding both passenger and freight trains: "It would be a tragedy if provision were not made for rail transportation on the proposed bridge over the Hudson at Fort Washington. In fact I will go so far as to say that neglect[ing] to seize this remarkable opportunity to afford means for the future binding together . . . of the two halves of the Metropolitan Region, now severed by one of Nature's greatest obstacles to intercommunication, for the sake of saving 3% of the cost, would be a blunder of the first magnitude. . . . A door half-open would indeed be a grave reflection on our vision and far-sightedness."[35]

A few days later, Adams again wrote to Wilgus and posed the essential question raised by the Port Authority, "whether or not the additional expenditure required to permit the later addition of rapid-transit tracks will be justified." Adams referred to Wilgus's previous letters and reports, in which he had "strongly advocated that the bridge should contain provision for rapid transit tracks."[36] Amman's initial plans included provisions for a lower deck to be added at some time in the future but with only two or three tracks for passenger trains and no freight tracks.

Adams met with Amman and other Port Authority officials and communicated to them Wilgus's insistence that the bridge also include freight tracks. Wilgus viewed a new bridge as part of the plan he had long advocated for a belt line railroad circling the entire port of New York. A rail link across the Hudson on the new bridge formed an essential component of such a belt line if the belt line then continued across the narrow upper section of Manhattan Island and over the Harlem River to the Bronx.

Amman, speaking for the Port Authority, wrote a letter to Adams commenting on Wilgus's insistence on strengthening the bridge for the addition of railroad tracks in the future. "I greatly appreciate the careful attention which you and Col. Wilgus have given to this question," he wrote, but then added a careful response: "I may assure you that your valuable suggestion will receive due consideration."[37] Amman explained that the Port Authority stood ready to make provision for as many as four "rapid transit" tracks to handle passenger trains but reserved judgment on adding even more strengthening of the bridge design for the much heavier freight trains.

Despite the insistence of Wilgus and the Regional Plan Association on provision for railroad tracks in the bridge design, the Port Authority received strong support from civic groups and politicians on both sides of the Hudson to build what would become the George Washington Bridge as proposed. An article in the *New York Times* on December 3, 1926, on the authority's last public hearing reported "unanimous approval of more than 200 civic and business organizations." The *Times* article mentioned that a future lower deck on the bridge "is to provide for four lines of passenger [train] traffic."[38] Adams clipped the article and sent it to Wilgus with a note suggesting that since the Port Authority seemed committed to a provision for passenger trains, the Regional Plan Association's goal had been achieved.

Wilgus could not hide his disappointment and urged the association to insist that the Port Authority think about the future: "A great mistake will be made if in the construction of this bridge the needs of the future are not fully anticipated in respect to mass transportation beneath the street surface. . . . It seems to me that the Port Authority is very narrow visioned in not now inviting the creation of cross-river means of rail transportation."[39] Colonel Delano, chairman of the Regional Plan Association, sent a letter to the chairman of the Port Authority congratulating the agency on securing approval for the bridge. On the basis of Wilgus's arguments, Delano urged the authority to design the bridge to accommodate six if not eight railroad tracks in the future, insisting that "mass transportation to and from . . . New Jersey must be provided for."[40] No formal response from the Port Authority is included in Wilgus's files.

After his consulting for the Regional Plan Association ended and as the George Washington Bridge neared completion in 1930, Wilgus continued to advocate for railroad tracks on the bridge. He wrote to his friend Ole Singstad, the engineer who completed the Holland Tunnel and now worked for the Port Authority, to press the case for the railroad tracks. In July 1930 Singstad wrote back on official Port of New York Authority letterhead: "I have discussed the matter with Mr. Amman and he informs me that provision is made for adding a lower deck for rapid transit . . . to carry the heaviest type of rolling stock. . . . Mr. Amman is of the opinion that the bridge will have ample strength to carry two rapid transit tracks and two trunk line railroad tracks."[41] Singstad added that the authority estimated the cost of adding a lower deck at $15 million but that the estimate did not include the cost of rail lines from the Hackensack Meadows or a needed tunnel through Bergen Hill. Singstad ended his letter with a gracious invitation for Wilgus to lunch with Amman and him when Wilgus next came to the city. Despite this letter, there is no evidence that the Port Authority ever seriously considered including mass transit or rail lines on the lower level of the bridge.

The celebrations surrounding the official opening of the George Washington Bridge on October 25, 1931, drowned out Wilgus's pleas for adding railroad tracks. The *New York Times* hailed the new bridge as "an astounding span of steel and wire" and reminded readers that the bridge, "a dream of three-quarters of a century," was now "the longest span ever built—3,500 feet between the two bridge towers, about two-thirds of a mile."[42] A total of 56,312 vehicles crossed the bridge the first full day of operation.

One year after the opening, the Port Authority reported that a total of 5.5 million vehicles and 500,000 pedestrians had crossed the bridge in the preceding year. As important, bridge tolls had generated $1.2 million in excess of costs for debt financing and maintenance.[43] Even as the Great Depression tightened

The George Washington Bridge, 1933, at the time the longest suspension bridge in the world and built by the quasi-public Port Authority. In the foreground are the tracks of the New York Central Railroad, a private railroad company. From the 1920s on, the railroads could not compete with the publicly subsidized highway system, the automobile, and the truck.

its grip, the Port Authority reported sufficient revenue from its tunnel and bridges to "to withstand five more years of depression with no harmful effects."[44] The authority projected a reserve balance of over $15 million even if traffic did not increase over the next few years. The authority's projections proved accurate. In 1937 more than 20 million vehicles used the agency's bridges and tunnel, an all-time high, assuring the agency of sufficient revenue to cover all expenses and a handsome "surplus" to use for future expansion.[45]

The Ongoing Transportation Issues in New York

Toll revenue enabled the Port Authority to be self-supporting and thus ensured its independence. Since it financed construction by borrowing directly in the bond markets, the authority no longer needed to go hat in hand to the states of New York and New Jersey for financing, which would have involved the legislative process with all of its entangling politics. With the automobile age well under way, the number of cars and trucks using the tunnels and bridges allowed the agency not only to cover its existing financing but also to use the ever-increasing surplus for the construction of more bridges and tunnels, airports, bus terminals, and office buildings.

The Port Authority rose to be a colossus dominating the future of transportation in the port of New York. By the late 1930s the agency was no longer focusing on its original mandate to improve rail freight transportation in the city, instead devoting its energy to providing access for automobiles and trucks, and later for airplanes. Meanwhile, the railroads continued their death spiral. Wilgus's plea to plan imaginatively for the future and maintain a balanced transportation system with modern rail and highway infrastructure fell on deaf ears. In 1962 the Port Authority added a lower deck to the George Washington Bridge. No railroad tracks were included, just six more lanes for cars and trucks.

Wilgus realized that some of the advantages offered by the private automobile and truck could not be ignored. Yet, even when serving as chairman of the Board of Consulting Engineers, designing the longest vehicular tunnel in the world, he argued that the new Holland Tunnel constituted only one piece of the solution to the port's problems. For balance, freight tunnels under the Hudson had to follow. And as we have seen, while he was a consultant to the Regional Plan Association, he fought to have the George Washington Bridge include rail lines as part of his long-held plan for an outer belt line railroad encircling the entire port.

Time and again he sought cooperation from the private railroads serving the port. He believed that a rational planning process demonstrating long-term advantages would persuade the railroads. Nothing proved to be farther from reality. The railroads could not see into the future and refused to surrender what they perceived to be their short-term advantage and participate in far-reaching, innovative plans to improve the movement of freight in the port. On the horizon, for the railroad executives to see if they chose to, were the truck and private automobile, whose eventual monopoly of transportation in the city was enabled by agencies like the Port Authority and the Triborough Bridge and Tunnel Authority willing to spend the equivalent of billions of dollars to build a highway, bridge, and tunnel infrastructure. Even as bankruptcy loomed and the railroads collectively careened toward disaster, they refused to participate in rail innovations proposed by Wilgus and the few other engineers dedicated to maintaining a balanced transportation system in the port, including modern rail facilities.

Usually at this point in the story of the collapse of rail and mass transportation in New York, Robert Moses appears as the archvillain. Certainly, Moses and his myriad authorities focused exclusively on the highway and bridge infrastructure, but his efforts did not stand alone. The Port Authority and the state highway authorities in New York, New Jersey, and Connecticut all spent millions and millions exclusively to support the highway revolution. Each new bridge, tunnel, or highway was hailed as the key to reducing congestion and speeding the trucks and automobiles on to final destinations. With hindsight of course, each new means of road transportation eventually only created more traffic and greater congestion.

A direct consequence of the increase in highway facilities was a continuing decline in the quality of the city and region's absolutely vital commuter rail ser-

vice. With the railroads losing profitable freight business to the truck and high-way, the railroads came to regard their commuter service as a money-losing enterprise. Both the New York Central and the Pennsylvania railroads, at one time among the largest and most profitable businesses in the United States, came to regard their magnificent Manhattan passenger stations as white ele-phants and started to neglect even basic maintenance.

With the largest mass transit subway system in the world, the city of New York had elicited the envy of the world. Operated by private business, the subways were initially immensely profitable. However, generations of New York City may-ors and politicians guaranteed New Yorkers the continuation of the "five-cent fare" and then the "ten-cent fare," regardless of the increasing costs of maintain-ing and modernizing the system. Eventually, the private operators walked away, and the city was obliged to take over operation of the subways as a public service. The slow death of the subways continued after World War II and into the 1960s and '70s, when the entire system almost collapsed.

For the commuter railroads and the subway the only solution was for the gov-ernment to take over and, at great public expense, revitalize both. At the dawn of a new century, attention has returned to rebuilding the freight rail system in the New York region. While it is too late to save the piers and shipping activity in Manhattan and Brooklyn, a revived freight transportation system modeled after the one proposed by William Wilgus over a hundred years ago may prove to be another part of the continued rebirth of New York.

Not only does the rebirth of the port's freight system continue, so do plans for

Verrazano Narrows Bridge, with Staten Is-land on the left, Brooklyn to the right, the Upper Bay of New York harbor, and Man-hattan in the dis-tance. The bridge stands in the exact location for the planned but never con-structed Nar-rows Tunnel for freight and sub-way trains.

new railroad passenger tunnels under the Hudson River to New Jersey. With planning well under way by the Port Authority for a new passenger tunnel from 34th Street in Manhattan to North Bergen, New Jersey, the newly elected Republican governor of New Jersey, Chris Christie, stunned the region with his announcement in October 2010 regarding the new rail link. In an incredibly short-sighted decision, the governor pulled the state of New Jersey out of the project, which was designed to double the number of New Jersey Transit commuter trains to Manhattan each day to relieve the paralyzing congestion in the Holland and Lincoln tunnels and on the George Washington Bridge.

In response the Bloomberg administration began work on an alternative plan to have the Metropolitan Transportation Authority extend the No. 7 subway line under the Hudson to Secaucus Junction in New Jersey. If constructed, the new tunnel would extend the New York City subway outside of the city boundaries for the first time and provide New Jersey commuters with "direct access to Times Square, Grand Central terminal and Queens."[46]

No responsible transportation expert or elected official is proposing another vehicular bridge or tunnel over or under the Hudson. Two years earlier in 2008, the state of New York announced plans to replace the Tappan Zee Bridge, which crosses the Hudson River north of New York City between Westchester and Rockland Counties. While the new span would have to accommodate highway traffic, the state plans to include at least two tracks for commuter trains. The No. 7 subway to Secaucus and commuter rail across the Tappan Zee would continue a renewed emphasis on a railroad solution to transportation in the New York region.

Wilgus at Rest

Coincidently with the opening of the George Washington Bridge in 1931, Wilgus decided to retire at the age of sixty-five. After a career spanning forty-five years from the time he secured his first railroad employment for the Minnesota & Northwestern Railroad in 1885, he planned, he said, to "devote my remaining years to voluntary unpaid service in what I conceived to be the public interest." But he did not anticipate the devastating effect of the depression on his own finances, describing "the stock market's cataclysmic plunge" as having carried him "to sorry depths."[47] Although he and his second wife spent more time in Vermont at Iridge, their home on the Connecticut River, Wilgus continued to work, but at a much slower pace.

In 1933 he conceived of the idea for a parkway through the state of Vermont, then his place of residence, to provide employment in response to the Great Depression. He traveled with the governor of Vermont to Washington to lobby the National Park Service to support the plan for a 250-mile Green Mountain National Parkway. Planned as part of the federal work relief program, the proposal generated great controversy across Vermont, and in 1936 the voters of Vermont decisively rejected the plan for a parkway.

In 1934 a much more challenging opportunity called Wilgus back to New York City. He received an invitation from Mayor LaGuardia and William Hodson, the city's commissioner of welfare and chairman of the New York State Emergency Relief Administration, to take over as director of work relief in the city. Work relief, part of the New Deal, provided federal funding to local agencies across the country to put the unemployed back to work. Across America, millions of people found work building public facilities and parks, creating an enduring legacy.

Wilgus feared that the position would entail a great deal of politics, "foreign to my way of life." Promised a free hand by Hodson, Wilgus accepted the position and temporarily took up residence in a Manhattan hotel. In accepting the position as director, Wilgus put the massive scale of the agency's efforts into perspective: "In terms of numbers of men employed and in terms of the wages and materials there is no engineering job under way in America or in the world for that matter that equals the job of directing the works division."[48] Newspaper articles placed the number of people on the work relief payroll in New York City at 130,000. Even when Wilgus was the director of the beginning of the Grand Central project in 1903, the work force, including the private contractors, never approached even a fraction of the number of people employed on work relief in New York during the depression years.

From his first days on the job, political problems arose. When Wilgus arrived, the work relief program listed 120,000 people on the payroll and had a chaotic administrative organization which Wilgus labored to reorganize and streamline. More problematic, calls constantly came from city politicians, "high and low, in all too many instances requesting favors."[49] Favors included employment for relatives and friends or demands to start a project in a favored neighborhood. Privately, Wilgus identified one of the most egregious offenders: Mayor LaGuardia.

The final straw came in late 1934 with an unexpected investigation of his agency launched out of the blue by Lloyd Stryker, a city alderman determined to rout out suspected corruption and generate favorable publicity for Stryker, who aspired to be mayor of New York. The aldermanic committee issued a series of sensational charges and set up an investigation of Wilgus, his staff, and the relief projects under way. According to one headline, the committee charged that the agency spent $2,242 a week on supervisors and less than $1,900 on the workers' payroll for "mosquito control."[50] Commissioner Hodson stormed into a meeting of the investigating committee in City Hall and angrily denounced the investigation as a "witch hunt" and the accusations of mismanagement as "absolutely false." Newspapers reported Hodson to be so angry that "his voice could be heard in every corner of City Hall." Wilgus, also at the meeting, adamantly supported Hodson and, responding to one alderman's charges, answered, "I say you are wrong."[51] He added his outrage at the articles in the newspapers suggesting that his agency was incompetent. Throughout a long and distinguished career Wilgus had never tolerated any questioning of his competence, especially from politicians seeking publicity at his expense.

Wilgus likened Stryker's investigation to the "Russian OGPU" and with great satisfaction reported: "My entire organization came through this soul-searching experience with flying colors, its honesty found beyond reproach."[52] Despite successfully defending his agency's integrity, Wilgus could not endure the ongoing "political intrigues" and handed in his resignation in May 1935. To document his vindication, he retained in his memoirs an elaborate testimonial from his staff given at a farewell dinner and letters from various senior officials praising his work as director. Mayor LaGuardia sent a brief note: "We certainly went through a difficult period together. While you believed that I differed with you, it was only a different approach from our relative positions. You did a grand job."[53] Always the politician, LaGuardia justified his interference in Wilgus's work as merely a different perspective, not as overstepping bounds.

After resigning, Wilgus spent the following winter in Winter Park, Florida. The end of his term as director of work relief in New York ended his long career. After 1935, he returned to New York only for brief periods of time. No other opportunities arose for him to devote his relentless energy and imagination to solving the city and port's transportation problems.

At his country home in Vermont he seemed to find some peace and contentment. With the coming of World War II, he would be again asked to lend his expertise to transportation issues. The United States, for the second time in his lifetime, found itself fighting in Europe, with its armies in need of efficient transportation—a problem Wilgus had solved once before while serving with the Allied Expeditionary Forces during World War I.

In November 1940, the chairman of the National Resources Planning Board, Frederic A. Delano, invited Wilgus to serve as a consultant in the preparation of a national transportation plan to support the American war effort. The Department of the Interior had established the Resources Planning Board in 1933 as part of the New Deal to plan public works projects nationwide. Wilgus had worked with Delano, President Roosevelt's uncle, previously, when Delano directed the Regional Plan Association in New York City and published the first comprehensive plan for the New York metropolitan region in 1929. Drawing on his experience during World War I, Wilgus told the Resources Planning Board that the key issue would be the need to nationalize the country's railroads to avoid the chronic shipping delays to the east coast ports, especially New York.[54]

In 1942, Wilgus, at his home in Vermont, received an invitation from the army's chief of the Railroad Branch to come to Washington to help plan the supply of Allied forces when they invaded France. A Colonel Ross informed Wilgus that all of his work would remain top secret. Wilgus asked a number of questions: where the invasion would take place; the "intended number of combatants and the nature of their equipment; the number of enemy troops to be coped with; . . . and the extent to which the Americans could look to their allies for man-power and supplies"—all logical questions needed to develop a comprehensive plan. To Wilgus's astonishment, Colonel Ross replied "that all was up to me to decide on my own."[55] Wilgus plunged ahead and completed a report, dated April 12, 1942, only to be informed that he could not keep a copy of the report because it would remain classified as top secret by the War Department until the end of the war. He never learned the extent to which his report assisted Eisenhower's campaign in France and Germany.

CONCLUSION

WILLIAM WILGUS LEFT a detailed chronicle of his professional life for posterity. His voluminous papers fill over one hundred boxes in the manuscripts collection of the New York Public Library. He overlooked no memo, telegram, or newspaper clipping; his papers include correspondence, memos, and reports detailing all his engineering projects. In an unpublished autobiography of over 350 pages, he described a professional career spanning more than forty years, from his start as a fledgling railroad engineer in Minnesota in 1885 to his brilliant ideas for the new Grand Central Terminal, his time as the chief consulting engineer for the Holland Tunnel in 1919, and finally, his tenure in 1933 as head of the Office of Work Relief in New York City during the Great Depression.

Yet, in the Wilgus papers at the New York Public Library, almost none of the material involves his personal life. In his memoirs he mentions meeting his first wife, May Reed, in Minneapolis, where he began work for the railroads. Her brother and Wilgus's friend, Charles Reed, who was at the time beginning his career as an architect, went on to design a number of stations for the New York Central Railroad and became involved with the Grand Central project. Wilgus included Reed's design for a new Grand Central terminal building among the first set of proposals he sent to the railroad's senior management in 1902. The New York Central and the Vanderbilts rejected Reed's twelve-story building and instead selected Whitney Warren's design for Grand Central. Wilgus's memoirs contain numerous references to Reed's design work, but May Reed, Wilgus's wife, receives only a passing word. Her sudden death in 1918 occurred while Wilgus served in France with the Allied Expeditionary Forces. He did not return to the United States to attend the funeral.

Even more elusive are his children. May and William had two children: a daughter, Margaret, born in 1892 and a son, William Jr., born in 1898. The 1900 census finds the Wilgus family living at 129 Riverside Drive in Manhattan. By 1910 they had moved to Scarsdale, to a large home on Crane Road, in a very fashionable neighborhood, next door to May's brother, Charles.

A scandal unfolded in 1911 when his daughter, Margaret, eloped at the age of nineteen with Clarence Smith of White Plains, New York. An article in the *New York Times* reported that the couple had married secretly to "outwit the parents of the bride, who thought that their daughter was too young to wed." Smith, a student at Amherst, returned to Scarsdale when their secret marriage became public. The couple went to the Wilgus home on Crane Road, told of their marriage, and "were forgiven."[1]

Wilgus's son, who used the nickname Jack, led a complicated life. He became involved in bohemian circles in New York City and included among his correspondents James Oppenheim, a poet and writer who lived in Greenwich Village. Oppenheim achieved a modicum of renown in radical circles when he published a collection of poetry, *Songs for a New Age*, with Century in 1914 and then founded a literary magazine, *New Age*, in 1916 with Waldo Frank. Jack Wilgus published some of his own poetry in *New Age* and kept up a long correspondence with Oppenheim.

In late January 1927, the New York newspapers reported that William Jr. had attempted suicide and was in Bellevue Hospital. Jack was distraught at the recent suicide of his girlfriend, Vivienne Minor, who killed herself by taking poison at the Martha Washington Hotel. While in Bellevue, Jack "cried out several times for aid from his father" and threatened to make another attempt at suicide.[2] William Wilgus never commented publicly or in his autobiography about this incident, nor did he include any details of his son's travails in his private papers.

After his suicide attempt, Jack left New York in 1930 and moved to Claremont, New Hampshire, across the Connecticut River from his father's home in Ascutney, Vermont. In Claremont, he leased a broken-down farm, the 140 acre Ironwood Mountain Farm, from the Wells family and tried to continue his literary career. It is not clear whether he had any contact with his father while living in Claremont.

In the summer and fall of 1931 Jack wrote a series of letters to Oppenheim describing his new life as a "farmer." In a July letter he describes himself as "poverty-stricken" and notes that he has sent his second book manuscript, *The Machinist*, to Harcourt Brace. He includes an essay entitled "On Annihilation" and characterizes the essay as a piece that "just about sums up my radicalism."[3] He also tells Oppenheim about his efforts to join a radical local political group.

One intriguing letter to Oppenheim, written in November 1931, describes growing up in the Wilgus household. Jack says that as a boy, he was "always lonely." He recounts his earliest memories traveling in his father's private railroad car, provided by the New York Central, and describes family life as "ever on the go." He sees his father as a distant man: "His eyes remained through the ages, raised, cold, and sad. He was the saddest man in the world."[4]

Another letter in November 1931 again sheds light on his father's personality from his perspective: "I learned universal distrust and dislike, narrowness and irritability and also self-distrust from my father." And he adds a devastating aside about his mother: "I learned defeatism from my cowed mother." Sundays required silence at the table, where "mother sat with downcast eyes" while "Father would read aloud some bit from the paper . . . and expatiate at length on some outrage. . . . Immigration was ruining the country. . . . The politicians were criminals. . . . Roosevelt, Wilson, Bryan, who not, were betraying our heritage. . . . On and on he talked while we were imbued with universal hatred and mistrust and learned prejudice."[5]

In a later passage he seems to soften and describes his father as "erect, haughty, Frenchy in his manners with a voice capable of expressing emotions while his face betrayed no affect. Sunday dinners included notable guests served formally by servants with fancy silver implements. Afterwards in the Library the men talked of 'rock-ballast, suspension bridges . . . electrification of one division or another.'" Once the guests departed, William Wilgus found nothing but fault: "Then the company was gone. My father had no longer to smile and bow and speak with soft-voiced modesty. The door closed behind the last guest and father inevitably turned upon my mother with reproaches about the dinner, about this and that she said, about the conduct of we children. Doors were banged and he went off to his club, leaving mother dissolved in tears and remorse."[6]

This portrait of Wilgus's personal life is quite at odds with his perceived public and professional persona, a role in which he received many laudatory letters and testimonials from co-workers, company officials, mayors, and governors. In his own autobiography, Wilgus is silent about his relationship with May Reed Wilgus. One can read his entire memoir and not know of his two children.

Two years after May Reed Wilgus died unexpectedly in October of 1918, Wilgus married Gertrude Tobin, an Irish immigrant, by way of Newfoundland, Canada. His memoirs remain almost silent about his second wife as well, with only the most fleeting of references and absolutely no mention of the relationship between his children and Gertrude. Nor does Jack have anything to say on this subject in his letters to Oppenheim.

Jack Wilgus's correspondence with Oppenheim ends in 1932 just before Oppenheim's untimely death from lung cancer. In a last letter Jack talks about starting a literary magazine with Oppenheim and asks whether Oppenheim thinks that they will "be able to find enough left-wingers and realists and debunkers to fill the little sheet? . . . I thought of calling it the Tempo. Hammer and Sickle would fair, but it is too outright a suggestion of Communism."[7] It is only too easy to imagine what William Wilgus might have thought of his son's will-o'-the-wisp pipe dream.

The Country Squire in Vermont

In 1922, Wilgus began to purchase land along the Connecticut River in Ascutney, Vermont, to build a country retreat, Iridge. He purchased land listed in the property records as the "so-called Lewis Farm" in two parcels and in 1923 added additional acreage for a small reservoir to ensure a supply of running water. The name "Iridge" evoked the memory of the Wilgus family's ancestral home in England. Perhaps a degree of whimsy led to the choice of the name.

Far from Buffalo, New York, and his modest upbringing, Wilgus enjoyed the life of the country squire in then still rural Vermont and looked forward to a comfortable retirement. In 1931, he donated a large portion of his property along the Connecticut River to the state of Vermont for a public park. Wilgus State

Park, constructed by the Civilian Conservation Corps, stretches for a mile along the river and provides picnic, camping, and canoeing facilities. Tall fir trees filter the warm summer sun as visitors enjoy the river just south of the large brick house Wilgus constructed as a personal retreat and retirement home.

A local history of Weathersfield, Vermont, the township that includes Ascutney, includes a short section about Wilgus with a critical comment about his personality: "The Colonel built his large house (sometimes known locally as the 'White Elephant') on his property in 1922 and retired to the life of a country squire. His aloof attitude toward the people of the village kept him from becoming a popular member of the community. . . . Mr. Wilgus did not like the address on his mail. Ascutneyville probably seemed a bit rustic to him."[8] Wilgus used his influence to have the name of the village changed to Ascutney, as it remains today. While he devoted a chapter of his autobiography to friendships among his fellow members of the Century Club in New York, he included no mention of any friends or acquaintances in Vermont.

In 1947, when old age and infirmities exacted a toll, William and Gertrude Wilgus sold their house in Ascutney and moved to Claremont, New Hampshire, just across the Connecticut River. Settled into a large home downtown at 230 Broad Street, Wilgus lived out the last years of his life. Downtown Claremont, a fading mill town, retains little of its earlier prosperity, and 230 Broad Street, long ago converted to a small apartment building, is not in the best of repair.

Wilgus did not die a wealthy man despite his long career as a highly paid railroad executive and consultant. His salary as fifth vice president of the New York Central Railroad rose to $50,000 in 1903, placing him among the highest-paid executives in American business at the time. Like millions of others, his finances were devastated by the stock market crash and the Great Depression. When he died in 1949, his estate went through probate court in Sullivan County, New Hampshire. An inventory filed on November 18, 1949, listed his total wealth at less than $30,000, including his house on Broad Street in Claremont, which was valued at just $15,000. His estate included a modest $10,000 in U.S. government bonds; he owned no other bonds, stocks, or any other investments.[9]

In his will, he bequeathed modest, almost miserly sums to the two children of his daughter, Margaret Wilgus Smith, who was now deceased. To each he bequeathed $200. To his son, William Jr., he left just $300, the exact amount he left to his long-time housekeeper in Claremont, Lucy Hatch. All of his remaining assets Wilgus left to his second wife, Gertrude. He had set up a trust for her at the Peoples National Bank in Claremont to be used to support her for the rest of her life and "such portion thereof . . . [as] she desires to contribute to the welfare of my beloved sisters-in-law Elizabeth M. Tobin and Mary A. Tobin." He provided for his second wife and her sisters rather than for his own children or grandchildren. In the will his bitterness toward his children remains clear.

In contrast to the lack of information about his family and personal life in his personal papers, Wilgus devotes an entire chapter of his memoirs, written in 1948, the last year of his life, to "Club Memberships Fostering Friendships: ca. 1899–1948." He enumerates his long-time membership in numerous professional and social organizations. In the chapter opening he quotes Samuel Johnson's advice to James Boswell: "If a man does not make new acquaintances as he advances through life, he will soon find himself alone. A man, sir, should keep his friendship in constant repair."[10]

Wilgus most proudly listed his membership in the socially prominent Century Club in New York City, which he joined in 1902. He describes forty-five years as an active member in very glowing terms: "Friendships have been fostered and repaired . . . by a warm esteem for each other's personality and character."[11] Wilgus includes a three-page list of "Century Club Members known to me and in large part close personal friends." Referred to as "friends," the men listed were in fact professional acquaintances.

Professional Achievements

In his memoirs Wilgus focused on his professional accomplishments. He entitled the last chapter in his chronicle of a long and illustrious career "Glance at Life's Happiness in Retrospect."[12] He mused over his professional career, with no reference to his personal life. Judgment of the measure of his work, he wrote, would "be determined in the first instance by those of my peers who may read or dig into evidence underlying this record." Above all else, Wilgus valued the opinion of his professional peers. Many chapters in his autobiography end with an appendix in which he included copies of testimonials or letters of appreciation from the companies, commissions, and especially the engineers with whom he worked.

Briefly he chronicles his life's journey from his early years in Buffalo and, given his father's "modest means," his own modest formal education. He describes wanting to be an engineer from the age of ten; at thirty-three he became the chief engineer of the New York Central & Hudson River Railroad and, just four months later, vice president, a title that embodied a true position of leadership and power when conferred in 1901.

Wilgus characterizes his plan for completely rebuilding the Central's aging facilities at 42nd Street as a "transfiguration," using a term from the Bible to describe his ideas. His concept for a new terminal facility remade the "heart" of the greatest city in the world. No false modesty intrudes; at the end of his life he knew with certainty his achievement at Grand Central. Despite the absence of formal recognition from the New York Central, his fellow engineers held his work on Grand Central to be among the most important engineering achievements in New York City's history. It remains so to this very day.

In addition to his masterpiece at Grand Central, Wilgus takes credit for the evolution of the Port Authority and his important role in the building of the Holland Tunnel. He chronicles his work on the Holland Tunnel as part of his continued efforts "to bring about a permanent solution of the country's transportation problems." Late in his career he records his public service during the Depression as director of federal work relief for the city of New York.

Not all plans and projects succeeded. During his retirement in Vermont, he advocated building a Green Mountain national park and parkway from the Massachusetts border to Canada. Modeled after the Skyline Parkway in Virginia and intended to provide much-needed work in Vermont during the Depression, the plan was decisively rejected by the voters. Never one to back down or admit mistakes, he comments, "I received what I conceived as an undeserved drubbing."

The last page of his autobiography adds a reflective note: "I indulge myself in the hope that in their essentials the fruits of my endeavors to evoke and direct the great sources of power in nature for the use and convenience of Man shall not be found wanting."

Not content to end on a contemplative note, he quickly directs the reader to consult an attached four-page summary of his career. The list includes eleven awards for service and engineering excellence, including the prestigious Telford Gold Medal, awarded in 1911 by the Institute of Civil Engineers of Great Britain and the Légion d'Honneur from France for his service during World War I. Both the Stevens Institute of Technology and the University of Vermont awarded him honorary degrees. While Wilgus can be self-congratulatory, what truly impresses the reader is the list of professional accomplishments:

- Rehabilitation, improvement and expansion of the New York Central Railroad & Hudson River Railroad.
- Electrification of the New York Suburban Zone of the New York Central Railroad.
- Inception of the Grand Central Terminal, New York City, predicated on the utilization of "air rights."
- Inception of the method of subaqueous tunnel construction first used by the Michigan Central R.R. crossing of the Detroit River.
- Inception of the idea of a joint State and federal investigation of the transportation needs of the Port of New York, of which aftermath was the New York Port Authority.
- Laying the foundations of the Transportation Service of the American Expeditionary Forces in World War I.
- Activities, *pro bono publico*, in cooperation with others, in the protection of the Mall, Washington, D.C., Morningside Heights Park, New York City and High Bridge Aqueduct . . . also in the establishment of a National Parkway and Park in the Green Mountains of Vermont, not endorsed by popular vote.
- Valuation of many leading railroads of the country.
- Services as a consulting engineer . . . in a variety of matters dealing with railroad planning, construction, maintenance, operation and financing. . . .
- Services on behalf of public agencies, as Chairman of the Board of Consulting Engineers of the Holland Tunnel, and as Consulting Engineer for the City of New York on the Narrows Tunnel project . . . Director of the Works Division in the City of New York . . . Lecturer Army War College.
- Counsel rendered for the Regional Plan of New York and its Environs in transportation matters.[13]

Throughout a long and truly distinguished career, Wilgus's life work centered on solving the transportation problems in the port and city of New York. To further

Wilgus's list of "causes initiated, lost and won"

Lost	Won
1. St. Louis River Hydro-Electric Development	1. Rehabilitation of New York Central
2. Small Car Freight Subway—Manhattan—1908	2. Electrification of the New York Central
3. Greater N.Y. Belt Line—1906–1926	3. Grand Central Terminal 1899–1907
4. Green Mountain National Parkway—1930s	4. Buffalo Union Station 1906–1908
5. Solving the Transportation Problem	5. Detroit River Tunnel 1905–1910
6. Lower Bay (N.Y.) Causeway Connecting	6. Port of New York Authority 1908–1917
Sandy Hook–Brooklyn	7. Transportation Services AEF 1917–1918
	8. Florida Ship Canal 1936–1939
	9. Park Protections
	10. Reorganization "Work Relief"
	11. Curing a Military Defect 1919–1942

SOURCE: Wilgus, "Milestones in the Life of a Civil Engineer," 1948, Linda Hall Library of Science, Engineering, and Technology, Kansas City, MO, 361.

dramatize his efforts over time to improve transportation, he included a tally of "causes initiated, lost, and won" (see table). To support the accounting he included references to his records left to the Library of Congress, the New York Public Library, and the United Engineers Societies Library (New York City), and to his four books and numerous published articles. In his accounting, he won far more than he lost.

Missing from the "lost" list is the Inter-terminal Belt Line in Manhattan. The "Greater N.Y. Belt Line" in the loss column includes the Narrows Tunnel. "Curing a Military Defect" referred to his continued involvement with the War Department during World War II, when the army asked Wilgus to come to Washington and participate in a secret planning process for the supply system that supported the Allied troops as they battled across France and into Germany after D-Day.

New York's Enduring Transportation Challenges

Wilgus's professional life focused on solving the transportation problems in New York City, especially the centuries-old problem of efficiently moving people and goods on and off Manhattan Island. Early in his career, at the age of thirty-three, Wilgus proposed the most complex transportation project in New York's history. His radical plan to electrify trains and then construct a two-story underground train yard in the heart of the busiest city in the country succeeded brilliantly. As a first step, the Grand Central project dramatically improved passenger access to Manhattan. With the opening of Pennsylvania Station by the Pennsylvania Railroad, New York now had the two most efficient and magnificent passenger rail facilities in the world.

Wilgus never succeeded in solving Manhattan's freight problems despite continuing to advocate for freight rail tunnels to link New York to the mainland. When serving as chair of the Board of Consulting Engineers planning the Holland Tunnel, he argued against letting the "trucking interests" take over the vehicular tunnel to haul freight under the Hudson. Even with the automobile and truck age fully under way, he knew the railroad offered vast efficiencies for hauling freight in the port of New York.

The cumbersome system of lighters, barges, and ferries used to carry freight across the harbor could have been improved with a careful planning process and the cooperation of the railroads serving the port. His advocacy for a Port Authority reflected his faith in a rational planning model. On the other hand, neither Wilgus nor the Port Authority could get the recalcitrant railroads to cooperate. In the end the railroads all went bankrupt; they never agreed to cooperate, nor did they recognize the threat looming from the trucks and passengers cars that soon flooded the publicly financed highways. How could a 40–horsepower truck traveling 20 miles an hour threaten the mighty railroads? With steam, electric, and soon diesel railroad engines capable of generating thousands of horsepower hauling long strings of freight cars, what other form of transportation could challenge the railroads? Eventually, bankruptcy came for the railroads, and the highway age ensued.

Ever since Henry Hudson's arrival in 1609, the unique geography of the port of New York has created daunting challenges. How to move people and commerce across the numerous waterways separating the various parts of the port remains a complex task. Despite the construction of bridges, highways and tunnels, Manhattan is still a narrow island surrounded by water.

In its *2005–2030 Transportation Plan*, the New York Metropolitan Transportation Council pointed out how important the Manhattan Central Business District, bordered by 60th Street on the north and the Battery on the south, remains after 400 years of settlement. On an average business day 3.7 to 3.8 million people enter and leave the business district, traveling between the "mainland" and Manhattan.[14] This is the equivalent of Connecticut's entire population (just over 3.5 million) plus another quarter of a million people. This army of commuters no longer work on the waterfront or labor in the factories or garment sweatshops; instead, they fill the office towers performing the work of a modern society.

Manhattan Island and the port of New York grew to become the most densely populated and busiest waterfront in the world. The city's rise to dominance began on the waterfront, and centuries of prosperity rested on the thriving port of New York and a once flourishing manufacturing sector. New York is no longer a manufacturing center. Today, finance, advertising, the media, and a host of businesses of a post-industrial society drive New York's economy. Legions work in service industries catering to the city's management class. Despite Manhattan's relentless pace of change, the island's narrow geography guarantees that transportation challenges remain paramount. Millions of people and thousands of tons of freight must move back and forth each day and night.

William Wilgus labored to overcome the city's and the port's geography. He used his relentless imagination to improve transportation in the port of New York, always believing railroad technology provided the key to moving people and goods over and under the port's rivers and bays. When he began working in New York City in 1899, the railroads still dominated the city and region's transportation. The age of the internal combustion engine had hardly begun. The German inventors Daimler, Otto, and Benz competed to patent early versions of the engine soon to revolutionize transportation. Daimler and Benz would go on to form a company to manufacture automobiles and named their famous cars after Benz's daughter, Mercedes. In 1900 Rudolf Diesel demonstrated his compression engine—soon to be universally referred to as the "diesel" engine—at the World Exposition in Paris.

By 1912, just a decade after Wilgus's arrival in New York and the start of the Grand Central project, the *New York Times* reported that motor vehicle registrations in the state of New York exceeded 85,000, over 10 percent of all vehicles registered in the United States. New York City led the state with 20,705 regis-

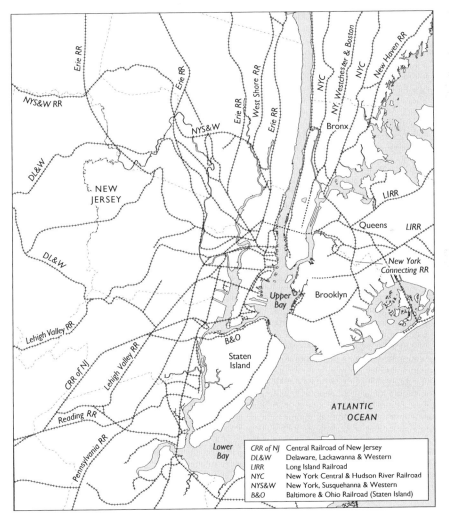

The culmination of the railroad era: rail lines throughout the port of New York, 1917.

CRR of NJ	Central Railroad of New Jersey
DL&W	Delaware, Lackawanna & Western
LIRR	Long Island Railroad
NYC	New York Central & Hudson River Railroad
NYS&W	New York, Susquehanna & Western
B&O	Baltimore & Ohio Railroad (Staten Island)

tered vehicles. More than 15,000 people possessed "chauffeur" licenses.[15] Wealthy New Yorkers who purchased fancy passenger automobiles did not learn to drive. Rather, their chauffeurs drove them about the city and out to Long Island or to Westchester County to their country homes and estates. As importantly, many freight trucks now carried goods and materials for short hauls through the crowded city streets, competing with the army of horse-drawn drays.

By 1919, the total number of registered vehicles in New York increased to 658,165, a 674 percent increase over 1912! Commercial trucks accounted for 51,527 of the vehicles registered in the New York City region and hauled more and more of the freight moved around the port.[16]

Along with the ever-increasing number of motor vehicles came demands for more roads and highways. Manhattan's grid pattern of streets, laid out in 1811, had established a uniform street width: most cross streets measure 60 feet, and the major north-south avenues stretch 100 feet from building to building. Never imagined to accommodate more than horse-drawn traffic, the streets barely handle the huge crush of cars and trucks today. Even with one-way traffic, the side streets, lined with parked cars on both sides, allow one lane of traffic to proceed at a snail's pace. Thousands of freight trucks double-park each day delivering the daily needs of the city. In spite of traffic laws and a small army of traffic officers roaming the streets to hand out tickets for violations, traffic often approaches gridlock. In fact, the news media periodically announce a "gridlock alert" for Manhattan whenever the president of the United States visits the United Nations or at the start of a major holiday.

Stand on any busy street corner in Manhattan and observe the absolute failure to develop any rational system for freight delivery or moving people on the surface of the city's streets. Without the subway, traveling about the city would be impossible. Buses and cars could handle only a small fraction of subway riders, and the elevated railroads are long gone, never to return.

Both the Port Authority and Robert Moses played key roles in the motor vehicle age and the eventual domination of car and truck in the region's subsequent transportation history. In his Pulitzer-winning biography, Robert Caro details Moses's rise to power in the 1930s and his push to construct highways and bridges in the metropolitan region at the expense of mass transit and the people.[17] Caro's assessment appears in the subtitle: "The Fall of New York." Yet long before Moses began his career, the pressure to construct highways and bridges in the port of New York had intensified. In 1910, the New York State Legislature established the Commission of Highways to improve roads throughout the state. Each year the commission pressed for increased funding, advocating for more limited-access highways to improve vehicular transportation. Highways and parkways followed.

For decades Wilgus and others advocated for an elevated railroad down the west side of Manhattan Island to serve the freight needs of the Hudson River piers and lower Manhattan. Part of the planning involved removing the New York Central's tracks from the city streets to eliminate "Death Alley." Unable to gain the cooperation of the Central, the city of New York eventually built an elevated roadway down the west side. In 1926, Manhattan Borough president Julius Miller proposed spending $11 million on the highway, and after heated political battles, including with the new Port Authority, the city decided to proceed. Construction of the city-financed highway began in 1929; when completed, the highway ran from 72nd Street to the Battery Park Tunnel. The West Side, one of the first urban highways in the country, further enabled the auto age.

With completion of the West Side Highway, the demand for more highways, tunnels, and bridges only accelerated. Moses built the Triborough Bridge, the Whitestone and Throngs Neck bridges, the Belt Parkway, the Cross Bronx Expressway—the list goes on and on. Nothing slowed the building of more highways and bridges for the next four or five decades. Eventually the web of highways mirrored the railroad network that once served the port of New York.

Finally the Port Authority, at great expense, relocated the port's shipping to Newark Bay, abandoning the hundreds of piers lining the Hudson and East rivers and the Brooklyn waterfront. The Port Authority spent millions of dollars to construct modern pier facilities in Newark and Elizabethport capable of handling a huge volume of shipping and easily adapted to the container revolution. Ironically, the Newark Bay piers include direct rail access enabling containers to be loaded directly onto railroad freight cars and hauled away—but not hauled by rail to New York City; only trucks carry freight onto Manhattan Island.

Back in Manhattan and Brooklyn, death came to the waterfront. Over time, the old piers disappeared, the West Side Highway came down, and New Yorkers reclaimed a waterfront that for over 300 years had provided prosperity for the city of New York. South of the Brooklyn Bridge on the Manhattan side of the East River, the South Street Seaport replaced piers where the *Flying Cloud* had once departed for San Francisco during the Gold Rush and where hundreds of ships had crowded the piers. Where tons of freight once moved each day, pedestrians now enjoy a walk along the shorelines of both Manhattan and Brooklyn.

New York is in the midst of another of its periodic metamorphoses, leaving behind a long history of shipping and industry to emerge in the twenty-first century as a vibrant post-industrial city. Many of the old industrial neighborhoods in Manhattan and Brooklyn now offer loft living in the spaces that once served a commercial and industrial economy. Young people fill the old immigrant neighborhoods and have reclaimed the city's waterfront.

As with all change, the present seems inevitable. William Wilgus recognized the lure of the private automobile and the flexibility of the individually operated truck. But he remained focused on the inherent efficiency of the railroad for freight transportation. After the expenditure of billions of dollars for highways in the port of New York, and across the country, a vision of a modern transportation system has turned again to the railroad as an essential part of the solution. In New York construction is under way for a commuter rail link from Long Island to Grand Central. Discussions and debate continue about building a railroad freight tunnel under New York harbor. A hundred years ago William Wilgus proposed the very same plans.

In a modest obituary published on October 25, 1948, the *New York Times* referred to Wilgus as a "rail expert" and mentioned his work on both Grand Central and the Holland Tunnel. The *Times* referred to his inspired railroad achievements and his participation in the construction of the world's longest vehicular tunnel at the dawn of the auto age. Simple words end his obituary: "a man of many ideas and causes."[18] Yet the *Times* placed his obituary on page 27 of the paper, far from the headlines of the day. The engineer who changed the face of midtown Manhattan and continued to dream large throughout a long and productive career deserved more. Wilgus asked what judgment history would make of his work. Stand in the magnificent Grand Concourse at 42nd Street in the heart of Grand Central, and the question is answered.

NOTES

Abbreviations

NYT *New York Times*

WJW William J. Wilgus

WJW, "Milestones" William J. Wilgus, "Milestones in the Life of a Civil Engineer," 1948, Linda Hall Library of Science, Engineering, and Technology, Kansas City, MO

WWP William J. Wilgus Papers, Manuscripts Division, New York Public Library

Introduction

1 "Solving Greatest Terminal Problem of the Age," *NYT*, Feb. 2, 1913, sec. 9, 2.
2 "A Glory of the Metropolis," *NYT*, Feb. 2, 1913, 16.
3 William J. Wilgus, "Milestones in the Life of a Civil Engineer," unpublished manuscript, 1948, Linda Hall Library of Science, Engineering, and Technology, Kansas City, MO, 10; hereafter cited as WJW, "Milestones."
4 Ibid., 18.
5 Ibid.
6 Ibid., 34.
7 Chauncey M. Depew, *1795–1895: One Hundred Years of American Commerce* (New York: D. O. Haynes, 1970), 111.
8 Kenneth T. Jackson, ed., *Encyclopedia of the City of New York* (New Haven: Yale University Press, 1995), xi.
9 U.S. Census Bureau, *Census Reports, 1900*, Vol. 7, *Manufactures, Part 1* (Washington, DC: Government Printing Office, 1902), 992, 998.

10 Russell Shorto, *The Island at the Center of the World* (New York: Doubleday, 2004).

Chapter 1. New York City's Geography and Transportation Challenges

1 Robert Albion, *The Rise of the Port of New York* (New York: Charles Scribner's Sons, 1939), 29.
2 Ibid., 62.
3 See Edwin Burrows and Mike Wallace, *Gotham: A History of New York City to 1898* (New York: Oxford University Press, 1999), pt. 1.
4 Ibid., 72–73.
5 Emory Johnson et al., *History of Domestic and Foreign Commerce of the United States* (Washington, DC: Carnegie Institution of Washington, 1915), 20.
6 G. W. Sheldon, "The Old Ship-Builders of New York," *Harper's New Monthly Magazine* 65, no. 386 (1884): 224–25.
7 Albion, *Port of New York*, 40–44.
8 Burrows and Wallace, *Gotham*, 430.
9 Albion, *Port of New York*, 88.
10 J. H. French, ed., *Gazetteer of the State of New York* (Syracuse, NY: R. P. Smith, 1860), 57.
11 William Cronon, *Nature's Metropolis: Chicago and the Great West* (New York: W. W. Norton, 1991), 43.
12 Ronald W. Filante, "A Note on the Economic Viability of the Erie Canal, 1825–1860," *Business History Review* 48, no. 1 (1974): 100.
13 Burrows and Wallace, *Gotham*, 653.
14 Albion, *Port of New York*, 400, app. 9.
15 Ronald Bailey, "The Other Side of Slavery: Black

Labor, Cotton, and Textile Industrialization in Great Britain and the United States," *Agricultural History* 68, no. 244 (1994): 44.

16 Albion, *Port of New York*, 101.

17 See Daniel Butler, *The Age of Cunard* (Annapolis, MD: Lighthouse Press, 2004).

18 See Edward J. Renehan, Jr., *Commodore: The Life of Cornelius Vanderbilt* (New York: Basic Books, 2007), chaps. 3–4.

19 Wheaton J. Lane, *Commodore Vanderbilt* (New York: A. A. Knopf, 1942), 184.

20 Lane, *Commodore Vanderbilt*, 185.

21 The New York Central began as a series of small independent railroads in upstate New York, the first of which, the Albany & Schenectady, began operations in 1831. The other railroads and their opening dates: Utica & Schenectady (1835), Syracuse & Utica (1839), Schenectady & Troy (1843), Rochester & Syracuse (1850), Rochester, Lockport & Niagara Falls (1852), Rochester & Buffalo (1852). Additional railroads also become part of the Central.

22 Roger Daniels, *Coming to America: A History of Immigration and Ethnicity in American Life* (New York: HarperCollins, 1990), 124.

23 U.S. Census Bureau, *Population of the United States in 1860* (Washington, DC: Government Printing Office, 1864), 609.

24 Frederick Knapp, *Immigration and the Commissioners of Emigration of the State of New York* (1870; New York: Arno Press, 1969), 37.

25 Ancestry.com, New York Passenger Lists, 1820–1957 (online database), September 1853; original data source: Passenger Lists of Vessels Arriving at New York, New York, 1820–1897, National Archives Microfilm Publication M237, Records of the U.S. Customs Service, Record Group 36, National Archives, Washington, DC.

26 French, *Gazetteer of the State of New York*, 109.

27 U.S. Census *Bureau, Report on the Manufactures of the United States in 1860* (Washington, DC: Government Printing Office, 1864), 384, 374.

28 U.S. Census Bureau, *Report on the Manufactures of the United States at the Tenth Census, June 1, 1880* (Washington, DC: Government Printing Office, 1883), 379–80, table 6.

29 U.S. Census Bureau, *Population of the United States at the Tenth Census, June 1, 1880* (Washington, DC: Government Printing Office, 1883), 670–71, table 25.

30 Walter Dodsworth, ed., *The Commercial Year Book: A Statistical Annual 1901* (New York: Journal of Commerce and Commercial Bulletin, 1901), 288.

31 Ibid.

32 Ibid., 294–97.

33 New York, New Jersey Port and Harbor Development Commission, *Joint Report with Comprehensive Plan and Recommendations* (Albany, NY: J. B. Lyon, 1920), 181–82.

34 "Port of New York Gained in Trade," *Greater New York* 10 (Feb. 7, 1921): 16.

35 Port and Harbor Development Commission, *Joint Report*, 188.

36 Ibid., 128–30.

37 Ibid., 128.

38 See Jill Jones, *Conquering Gotham: A Gilded Age Epic; The Construction of Penn Station and Its Tunnels* (New York: Viking, 2007).

Chapter 2. The Brilliance of Grand Central

1 WJW, "Milestones," chap. 5, "Transition East to Upstate New York: 1893–1897."

2 Ibid., 74.

3 Ibid., 79.

4 Carl W. Condit, *The Port of New York: A History of the Rail and Terminal System from the Grand Central Electrification to the Present* (Chicago: University of Chicago Press, 1981), 277, table 7.

5 "The Grand Central Station" (editorial), *NYT*, Mar. 26, 1899, 18.

6 Editorial, *NYT*, Sept. 20, 1899, p. 6, col. 5.

7 "Fifteen Killed in Rear End Collision," *NYT*, Jan. 9, 1902, 1.

8 WJW, "Milestones," 91.

9 WJW, "The Grand Central Terminal in Perspective," *Transactions of the American Society of Civil Engineers* 106 (Oct. 1940): 197.

10 Ibid., 1003.

11 WJW to William H. Newman, Dec. 22, 1902, WWP, box 1.

12 WJW to Newman, Mar. 19, 1903, WWP, box 1.

13 "N.Y. Central's Report," *NYT*, Oct. 4, 1902, 5.

14 WJW, "Milestones," 93.

15 WJW, "Grand Central Terminal in Perspective," 1003.

16 "The New York Central's Interest Free Terminal," *Railway Age Gazette* 52, no. 11 (Mar. 15, 1912): 462.

17 See David Nye, *Electrifying America: Social Meanings of a New Technology, 1880–1940* (Cambridge, MA: MIT Press, 1990).

18 Frank J. Sprague, "The Electric Railway," *Century Magazine* 70 (Aug. 1905): 512.

19 Bion Arnold, "Report of Bion T. Arnold to William J. Wilgus upon the Proposed Electric Equipment of the Hudson Division of the New York Central Railroad," Bion T. Arnold Papers, Manuscripts Division, New York Public Library, box 6.

20 Bion Arnold, "A Comparison Study of Steam and Electric Power for Heavy Railroad Service," *Railroad Gazette* 34, no. 26 (June 27, 1902): 498.

21 WJW, Meeting Minutes, WWP, box 6, Electric Traction Commission, 1902–1906.

22 The Harlem Division ran from Grand Central to White Plains and then north through Putnam and Columbia Counties to Chatham, New York; it was originally the New York & Harlem Railroad. The Hudson Division ran from Grand Central through the southern Bronx and then along the Hudson River to Croton-on-Hudson, Poughkeepsie, and then north to East Albany; it was originally the Hudson River Railroad.

23 "The New York Central's Terminal Electrification at New York," *Railroad Gazette* 41, no. 14 (Oct. 5, 1906): 293.

24 WJW, "Milestones," 101.

25 WJW, "Grand Central Terminal in Perspective," 1007.

26 Sprague, "Electric Railway," 522.

27 Arnold, "Report of Arnold to Wilgus upon the Proposed Electric Equipment," 7.

28 George Westinghouse to W. H. Newman, President, New York Central & Hudson River R.R. Co., *Railroad Gazette* 39, no. 25 (Dec. 22, 1905): 579–80.

29 Frank J. Sprague, "An Open Letter to Mr. Westinghouse," *Railroad Gazette* 40, no. 1 (Jan. 1906): 8.

30 WJW, "The Electrification of the Suburban Zone of the New York Central and Hudson River Railroad in the Vicinity of New York City," *Transactions of the American Society of Civil Engineers* 51, no. 1079 (Mar. 18, 1908): 77.

31 "Electrification of the New York Central in and near New York City," *Railway Age*, Jan. 26, 1908, 145.

32 WJW, "Milestones," 111.

33 Minutes of Meetings of the Electric Traction Commission, May 19, 1903, WWP, box 6.

34 "Test of the New York Central Electric Locomotives," *Railroad Gazette* 37, no. 23 (Nov. 18, 1904): 552.

35 Ibid., 554.

36 "Electric Engine Beats All Rivals," *New York Herald*, Nov. 13, 1904, 4.

37 "First Electric Train Enters New York City, Thousands of Spectators Cheer," *World*, n.d.

38 WJW, "Milestones," 115.

39 Nye, *Electrifying America*, 27.

40 WJW, "Record of the Inception and Creation of Grand Central Terminal Improvement, 1902–1913," WWP, box. 1.

41 "Twenty-Five Killed and Seventy-Five Injured," *New York Herald*, Feb. 17, 1907, 1.

42 WJW, "Milestones," 118.

43 Ibid., 119.

44 WJW, "Woodlawn Wreck," WWP, box 7, 5.

45 WJW, "Milestones," 119.

46 Ibid., 120.

47 Ibid., 122.

48 "Erecting the Grand Central Terminal," *Engineering Record* 66, no. 8 (Aug. 24, 1912): 222.

49 Hugh Thompson, "The Greatest Railroad Terminal in the World," *Munsey's Magazine* 45 (April 1911): 50.

50 "New Grand Central Terminal Opens Its Doors," *NYT*, Feb. 2, 1913, sec. 9, 1.

51 *Poor's Manual of Railroads* (New York: H. V. & H. W. Poor, 1915), 83.

52 WJW, "The Grand Central Terminal in Perspective," 1016.

53 Ibid.

54 Ibid., 1021.

55 WJW, "Inception and Creation of Grand Central Terminal Improvement," WWP, box 1.

56 Bernard Walker, "The World's Greatest Railway Terminal," *Scientific American* 104, no. 25 (June 17, 1911): 594.

57 WJW, "Milestones," 166.

58 Ibid., 166.

Chapter 3. New York's Freight Problem

1 See New York State, Laws of 1846, chap. 216.

2 *Report of the Commission to Investigate the Surface Railroad Situation in the City of New York on the West Side* (Albany, NY: 1918), 22.

3 See New York State, Laws of 1869, chap. 917.

4 *Report of the Commission to Investigate,* 26.

5 New York, New Jersey Port and Harbor Development Commission, *Joint Report with Comprehensive Plan and Recommendations* (Albany, NY: J. B. Lyon, 1920), 151.

6 *Report of the Commission to Investigate,* 12.

7 New York State, Laws of 1906, chap. 109.

8 *Report of the Commission to Investigate,* 202–4.

9 WJW, "Plan of Proposed New Railroad System for the Transportation and Distribution of Freight by Improved Methods," 12, WWP, box 45.

10 WJW, "Analysis of Various Plans for Improved Freight Facilities in Manhattan," WWP, box 46, Small Car Freight Subway.

11 See James B. Walker, *Fifty Years of Rapid Transit* (New York: Law Printing, 1918).

12 William Willcox to WJW, May 2, 1908, WWP, box 45, Correspondence.

13 WJW to Edward Bassett, Public Service Commission, 4/17/1908, WWP, box 45, Correspondence.

14 WJW, "Date Required in Connection with Manhattan Island Freight Report," WWP, box 45, Transfer Case 2.

15 WJW to Percy R. Todd, May 5, 1908, WWP, box 45, Transfer Case 2.

16 Todd to WJW, May 7, 1908, WWP, box 45, Transfer Case 2.

17 WJW to Todd, May 8, 1908, WWP, box 45, Transfer Case 2.

18 WJW to Edward Bassett, Public Service Commission, May 11, 1908, WWP, box 45, Transfer Case 2.

19 "Brief Submitted on Behalf of the Amsterdam Corporation by William J. Wilgus, to the Public Service Commission for the First District," Dec. 26, 1908, WWP, box 45.

20 See Keith Revell, *Building Gotham: Civic Culture and Public Policy in New York City, 1898–1938* (Baltimore: Johns Hopkins University Press, 2003), 2–14.

21 "Gigantic Plan to Relieve Street Congestion," *NYT,* Oct. 4, 1908, Sunday Magazine, 10.

22 Francis Lane, "A Proposed Subway Belt Line for Lower New York," *Engineering News* 60, no. 16 (Oct. 15, 1908): 403–5.

23 "A Plan for Subway Freight Distribution in New York" (editorial), *Engineering News* 60, no. 16 (Oct. 15, 1908): 419.

24 Charles W. Baker to WJW, Oct. 19, 1908, WWP, box 45, Transfer Case 2.

25 WJW, "Proposed New Freight Subway: New York City and Port." *Railway Age Gazette* 45, no. 20 (Oct. 16, 1908): 1150–57.

26 "Truckmen to Fight Freight Subway," *NYT,* Nov. 8, 1908, 8.

27 W. H. Truesdale to WJW, July 20, 1908, WWP, box 45, Transfer Case 2.

28 WJW to Truesdale, July 21, 1908, WWP, box 45, Transfer Case 2.

29 WJW to W. H. Newman, June 23, 1908, WWP, box 45, Transfer Case 2.

30 Henry Pierce to WJW, Sept. 7, 1908, WWP, box 45, Transfer Case 2.

31 WJW to Edward Bassett, Sept. 10, 1908, WWP, box 45, Transfer Case 2.

32 Bassett to WJW, Sept. 17, 1908, WWP, box 45, Transfer Case 2.

33 E. V. W Rooseter to WJW, Oct. 5, 1908, WWP, box 45, Transfer Case 2.

34 See Revell, *Building Gotham,* 105–14.

35 WJW, "Brief Submitted on Behalf of the Amsterdam Corporation," 18.

36 Ibid., 20.

37 WJW, "Proposed New Freight Subway," 1150–57.

38 WJW, "Memorandum," Nov. 11, 1914, WWP, box 45.

39 Ibid.

40 WJW, "Comments on Plan Proposed by Calvin Tomkins," July 22, 1910, WWP, box 56.

41 Report of the Commission to Investigate, 7.

42 Port and Harbor Development Commission, *Joint Report,* 151.

43 Friends of the High Line, "High Line: High Line History," High Line website, www.thehighline.org/about/highlinehistory.html.

Chapter 4. Expanding the Subway in Manhattan

1 Clifton Hood, *722 Miles: The Building of the Subways and How They Transformed New York* (New York: Simon and Schuster, 1992).

2 WJW to Public Service Commission, Mar. 9, 1909, WWP, box 47, folder A4.

3 "City Railways," *NYT Supplement,* Apr. 16, 1859, 2.

4 "Broadway," *New York Daily Times,* Apr. 19, 1852, 2.

5 "How to Get Down Town" (editorial), *New York Daily Times*, May 9, 1860, 4.

6 New York State, Laws of 1831, 263.

7 "Streetcars," *Encyclopedia of the City of New York*, ed. Kenneth Jackson (New Haven: Yale University Press, 1995), 1127.

8 See Robert Reed, *The New York Elevated* (South Brunswick, NJ: S. A. Barnes, 1978), and William F. Reeves, *The First Elevated Railroads in New York* (New York: New York Historical Society, 1936).

9 "The Elevated Railway," *NYT*, Sept. 7, 1869, 2.

10 "Completion of the Gilbert Elevated Railroad: The Station at Fourteenth Street," *Frank Leslie's Illustrated Newspaper*, June 8, 1878, 242.

11 Hewitt, quoted in Interborough Rapid Transit Company, *Interborough Rapid Transit: The New York Subway* (New York: Interborough Rapid Transit Co., 1904), 16.

12 "The City's Official Vote," *NYT*, Nov. 24, 1894, 2.

13 "Subway a Year Old: What It Has Done," *NYT*, Oct. 27, 1905, 9.

14 To provide a comparison, the Erie Railroad, one of the four major trunk railroads in the country, with 2,230 miles of tracks, provided service between New York and the Midwest. For the 1908–9 fiscal year, revenue totaled $50,441,161. This was twice the revenue of the IRT, but the IRT provided rapid transit only in New York City and carried no freight. With much higher expenses, the Erie covered its bond payments but paid no dividends on either preferred or common stock in 1908 or 1909. New York's first subway proved to be a much more profitable investment than even one of the major railroad companies in the United States.

15 See Hood, *722 Miles*, 121–32.

16 "Transit Board under Siege," *NYT*, Dec. 15, 1905, 18.

17 "One Subway, Belmont Declares," *NYT*, Jan. 12, 1908, 2.

18 "Private Capital for Subways," *NYT*, Apr. 23, 1908, 8.

19 "Routes Laid Out for New Subways: Relation the Tri-Borough System Will Bear to the Lines Now in Existence," *NYT*, Feb. 18, 1910, 9.

20 "Meeting to Discuss New York Subway Situation," *Electric Railway Journal* 36 (Oct. 22, 1910): 878.

21 "The Growing Problem of Lower Manhattan," *NYT*, Aug. 30, 1908, sec. SM3, 1.

22 WJW and the Inter-Terminal Belt Line to Public Service Commission, March 9, 1909, WWP, box 47, folder A4.

23 For biographical material on Schwab, Halsey, White, and Hodenpryl, see John Lenard, ed., *Who's Who of Finance, Banking and Insurance, 1920–1922* (Philadelphia: Rex Printing House, 1922).

24 "Memo to file: Phone interview with Commissioner Bassett, February 13, 1909," WWP, box 47, folder 40.

25 "Memo to file: Phone interview with Commissioner Bassett, February 19, 1909," WWP, box 47, folder 40.

26 Henry Pierce to WJW, Mar. 4, 1909, WWP, box 47, folder 40.

27 WJW to Public Service Commission, Mar. 9, 1909, WWP, box 47, folder A4, 7–10.

28 "(Exhibit 'A') Suggested Inter-Terminal Belt Line for Passengers in the Borough of Manhattan," WWP, box 47, folder A4, 1.

29 WJW to Public Service Commission, Mar. 9, 1909, Exhibit A, 3, WWP, box 47.

30 WJW to Public Service Commission, Mar. 9, 1909, WWP, box 47, folder A4, 5.

31 See Daniel R. Gallo and Frederick A. Kramer, *The Putnam Division: New York Central's Bygone Route through Westchester County* (New York: Quadrant Press, 1981).

32 *Travelers' Railway Guide, March 1903* (Chicago: American Railway Guide Co., 1903), 390.

33 Pierce to W. C. Brown, New York Central Railroad, Apr. 5, 1909, WWP, box 48, folder Inter-Terminal Belt Line.

34 *Report of the Public Service Commission*, vol. 2, *Statistics of Transportation Companies for the Year Ended June 30, 1914* (Albany, NY: J. B. Lyon, 1915), 18.

35 Hood, *722 Miles*, 159.

36 Public Service Commission, *Statistics of Transportation*, 2:19.

37 "Subway Schemes Multiply: Tube Elevated Plan of the Amsterdam Corporation," *Evening Sun*, Mar. 10, 1909.

38 "Mr. Wilgus's Rapid Transit Proposal" (editorial), *Railway Age Gazette* 47, no. 10 (Mar. 12, 1909): 512.

39 WJW to Editor, *New York Daily Tribune*, Mar. 13, 1909, WWP, box 47, Correspondence.

40 WJW to R. Ferris, Editor, *Railway Age Gazette*, Mar. 12, 1909, WWP, box 48, Correspondence.

41 W. C. Brown, New York Central, to Henry Pierce, Mar. 18, 1909, WWP, box 47, Correspondence.

42 Pierce to Brown, Mar. 22, 1909, WWP, box 47, Correspondence.

43 Brown to Pierce, Mar. 25, 1909, WWP, box 47, Correspondence.

44 Pierce to R. C. Shepard, Apr. 5,1909, WWP, box 47, Correspondence.

45 "Both Houses Pass New Transit Bill," *New York Herald*, Apr. 24, 1909.

46 Pierce to Ira Place, New York Central, Aug. 10, 1909, WWP, box 47, Correspondence.

47 Pierce to Shepard, Sept. 8, 1909, WWP, box 47, Correspondence.

48 "Mayor McClellan's Vetoes," *The World*, May 17, 1909.

49 "Get Axe Ready for 11th Avenue Tracks," *NYT*, May 29, 1909, 6.

50 Pierce to Thomas Cuyler et al., Sept. 18, 1909, WWP, box 47, Correspondence.

51 WJW to Inter-Terminal syndicate members, Sept. 20, 1909, WWP, box 47, Correspondence.

52 "Select a Committee of Subway Advisors: Chamber of Commerce and Merchants' Association Make Known Their Choice," *NYT*, Dec. 3, 1910, 4.

53 "One Subway Best Mayor's Board Finds," *NYT*, Dec. 29, 1910, 4.

54 Editorial, *New York Daily Tribune*, June 15, 1911.

55 Editorial, *Globe and Commercial Advertiser*, June 14, 1911.

56 "The Subway Report" (editorial), *NYT*, June 14, 1911, 8.

57 Hood, *722 Miles*, 159.

58 See Railroad.net, "New York Central's 1934 West Side Improvement," www.railroad.net/articles/railfanning/westside/index.php.

59 R. L. Duffus, "Vast Project to Relieve City's Traffic Jams," *NYT*, Mar. 17, 1929, XX3.

60 See Friends of the High Line, "High Line History," High Line website, www.thehighline.org/about/highlinehistory.html.

Chapter 5. World War and Ideas for a New York–New Jersey "Port Authority"

1 WJW to A. R. Smith, Secretary, Barge Canal Commission, Dec. 29, 1909, WWP, box 48, Movement Leading to N.Y.–N.J. Interstate Legislation Respecting a Port of New York Authority.

2 WJW, Memo, Dec. 29, 1909, p. 12, WWP, box 48, Movement Leading to N.Y.–N.J. Port Authority.

3 WJW, "A Suggestion for an Interstate Metropolitan District," 15–16, WWP, box 48.

4 Ibid., 16–17.

5 Ibid., 21.

6 WJW to A. R. Smith, Dec. 31, 1909, WWP, box 48, Movement Leading to N.Y.–N.J. Port Authority.

7 WJW to Gustav Schwab, Chairmen, Committee on Foreign Commerce and Revenue Laws, Chamber of Commerce, Jan. 11, 1910, WWP, box 48, Movement Leading to N.Y.–N.J. Port Authority.

8 WJW to Charles Whiting, Editor-in-Chief, *Engineering News*, Mar. 9, 1910, WWP, box 48, Movement Leading to N.Y.–N.J. Port Authority.

9 WJW, "The Cost of Freight Handling in the Port of New York." *Engineering News* 63 (Mar. 31, 1910): n.p.

10 A. R. Smith to WJW, Apr. 26, 1910, WWP, box 48, Movement Leading to N.Y.–N.J. Port Authority.

11 WJW to Hon. J. A. Johnson, Senate House, Trenton, N.J., Feb. 14, 1911, WWP, box 48, Movement Leading to N.Y.–N.J. Port Authority.

12 "New York's Freight Terminal Problem," *Railway Age Gazette* 54 (Apr. 11, 1913): 843.

13 WJW, "Comments on Plan Proposed by Calvin Tomkins, Commissioner of Docks, for a Joint Terminal on the North River," WWP, box 56.

14 WJW, "Preliminary Study for a Greater New York Belt Line," Oct. 7, 1913, WWP, box 55.

15 Ibid., 11.

16 WJW, "Memorandum of Agreement," WWP, box 55.

17 WJW to E. R. Thomas, President, Lehigh Valley Railroad, Oct. 7, 1913, WWP, box 48; Thomas to WJW, Oct. 16, 1913, WWP, box 48.

18 WJW to Samuel Rae, President, Pennsylvania Railroad, Nov. 10, 1913, WWP, box 48.

19 Rae to WJW, Nov. 19, 1913, WWP, box 48.

20 WJW to Henry Hodge, Public Service Commission, Dec. 22, 1916, WWP, box 56.

21 For a discussion of the New York Harbor case, see Jameson Doig, *Empire on the Hudson: Entrepreneurial Vision and Political Power at the Port of New York Authority* (New York: Columbia University Press, 2001), 28–30.

22 WJW to Henry Hodge, Jan. 9, 1917, WWP, box 56.

23 WJW to Hon. Charles S. Whitman, Governor of the State of New York, Jan. 19, 1917, WWP, box 56.

24 Doig, *Empire on the Hudson*, 41, and see 39–46 for further discussion of the New York Harbor case.

25 New York State, Laws of 1917, chap. 426, and State of New Jersey, Laws of 1917, chap. 130.

26 WJW, "Milestones," 181.

27 Ibid.

28 Ibid., 183.

29 WJW, *Transporting the A.E.F. in Western Europe: 1917-1919* (New York: Columbia University Press: 1931), 21.

30 WJW, "Milestones," 187.

31 Ibid.

32 Key chapters in Wilgus's *Transporting the A.E.F.* highlight the complexity of the transportation system:

Chapter 6 . . . . Headquarters Staff
Chapter 7 . . . . Engineering
Chapter 11 . . . . Traffic
Chapter 12 . . . . Lines of Communication
Chapter 13 . . . . Terminals, Sidings, and Multiple Tracking
Chapter 14 . . . . Locomotives and Cars
Chapter 17 . . . . Ports
Chapter 18 . . . . Cranes
Chapter 27 . . . . Ocean Shipping

33 WJW, *Transporting the A.E.F.*, 137.

34 Ibid., 313.

35 Ibid., 317.

36 All 1916 railroad data are from *Poor's Intermediate Manual of Railroads, 1917* (New York: Poor's Manual Company, 1917).

37 WJW, *Transporting the A.E.F.*, 344.

38 Ibid., 557.

39 WJW to General John J. Pershing, Commander-in-Chief, American Expeditionary Forces, France, Nov. 18, 1918, WWP, box 27, General Correspondence, 1917 Jan.–1918 Dec.

40 Pershing to WJW, Dec. 2, 1918, WWP, box 27, General Correspondence, 1917 Jan.–1918 Dec.

41 F. L. Whitley, Adjutant General, American Expeditionary Forces, to WJW, Mar. 12, 1919, box 27, General Correspondence, 1919 Jan.–1920 Dec.

42 WJW, "Milestones," 197

43 Ibid., 210.

44 Clifton Hood, *722 Miles: The Building of the Subways and How They Transformed New York* (New York: Simon & Schuster, 1992), 193–97.

45 Editorial, *Railway Age Gazette* 62 (May 11, 1917): 985–86.

46 "Appeal for 'Port Authority' Sent to Members," *Greater New York: Bulletin of the Merchant's Association of New York* 8 (Mar. 24, 1919): 18.

47 "Movement Leading to N.Y.–N.J. Port Authority," WWP, box 48.

48 Doig, *Empire on the Hudson*, 50–53.

49 New York, New Jersey Port and Harbor Development Commission, *Joint Report with Comprehensive Plan and Recommendations* (Albany, NY: J. B. Lyon, 1920), 57.

50 New Jersey Port and Harbor Development Commission, *Progress Report* (Albany, NY: J. B. Lyon, 1919), 25.

51 Ibid., 52.

52 Harbor Development Commission, *Joint Report*, 2–3.

53 Ibid., 118.

54 Ibid., 173–74.

55 Ibid., 251.

56 Ibid., 280.

57 New York State, Laws of 1921, chap. 154, and State of New Jersey, Laws of 1921, chap. 151. A copy of the compact is provided in the appendix to Doig, *Empire on the Hudson*, 403–9.

58 Sidney Goldstein, "An Authority in Action: An Account of the New York Authority and Its Recent Activities," *Law and Contemporary Problems* 26 (1961): 715.

59 Ibid., 719.

60 "Engineers Favor Better Port Plan," *NYT*, Feb. 8, 1921, E1.

61 Alexander Archibald (mayor of Newark), "Greater Newark in the Making," *Port of New York Harbor and Marine Review* 2 (Feb. 1922): 32.

62 "Movement Leading to N.Y.–N.J. Interstate Legislation Creating a Port of New York Authority," WWP, box 48.

63 WJW to Julius Henry Cohen, July 18, 1925, WWP, box 56.

64 Cohen to WJW, Aug. 7, 1925, WWP, box 56.

65 Ibid., 2.

66 Ibid., 1.

67 See Doig, *Empire on the Hudson*, 103–14.

68 Ibid., 105.

69 John Griffin, *The Port of New York* (New York: City College Press, 1959), 82.

70 Goldstein, "An Authority in Action," 720.

Chapter 6. Making Room for the Automobile

1 See B. H. M. Hewett and S. Johannesson, *Shield and Compressed Air Tunneling* (New York: McGraw Hill, 1922), 28–37.

2 "The Broadway Pneumatic Tunnel," *Frank Leslie's Illustrated Newspaper*, Feb. 19, 1870, 381.

3 For a comprehensive recounting of the struggles of the Pennsylvania Railroad to obtain a franchise for the Hudson and East River tunnels, see Jill Jones, *Conquering Gotham* (New York: Viking, 2007), 73–126; tunnel construction is described on pp. 151–232.

4 Hewett and Johannesson, *Shield and Compressed Air Tunneling*, 38, 40, 44.

5 Ibid., 2.

6 State of New York, *Report of the New York Interstate Bridge Commission*, Senate Report no. 24, Feb 8, 1907, 4.

7 Henry Petroski, *Engineers of Dreams* (New York: Alfred Knopf, 1995), 122–216.

8 Ibid., 138.

9 "G. Lindenthal Calls for Hudson Bridge," *NYT*, Dec. 9, 1912, 20.

10 Petroski, *Engineers of Dreams*, 212.

11 "Great Hudson Span Close to Reality," *NYT*, May 1, 1921, 36.

12 J. A. L. Waddell, "Bridge versus Tunnel for the Proposed Hudson River Crossing at New York City," *Transactions* 84, no. 1477 (1921): 570–74.

13 J. A. L. Waddell, *Economics of Bridgework* (New York: John Wiley, 1921).

14 Waddell, "Bridge Versus Tunnel," 571.

15 Ibid., 574.

16 "A Highway under the Hudson" (editorial), *Engineering News-Record* 84 (Feb. 19, 1920): 356.

17 New York State, Laws of 1919, chap. 178, 1.

18 WJW, "Milestones," 211.

19 "Duties of the Board of Consulting Engineers," WWP, box 49.

20 "Report of the Board of Consulting Engineers," Dec. 31, 1919, plate 29, WWP, box 49.

21 State of New York, *Report of the New York State Bridge and Tunnel Commission, Legislative Document* no. 64 (Albany, NY: J. B. Lyon, 1921), 18.

22 George H. Pride, "A Layman's Viewpoint of the Proposed Hudson Vehicular Tunnels," *Engineering News-Record*, Mar. 18, 1920, n.p.

23 WJW to C. M. Holland, Chief Engineer, N.Y.–N.J. Interstate Bridge and Tunnel Commission, Mar. 20, 1920, 1, WWP, box 49.

24 WJW, "Milestones," 151–52.

25 For details of the Detroit River construction see Charles Crandall and F. Barnes, *Railroad Construction* (New York: McGraw Hill, 1913), 120–23.

26 "The Detroit River Tunnel," *Railway Age Gazette* 51 (Nov. 10, 1911): 947.

27 WJW, "Milestones," 159.

28 Ibid., 162.

29 Ibid., 160.

30 *Report of the New York State Bridge and Tunnel Commission*, 25.

31 Ibid., 24.

32 Appendix 2: Cast Iron Tunnel Analysis, ibid., 53–89; quotation on p. 55.

33 *Report of the New York State Bridge and Tunnel Commission*, 96.

34 Ibid., 155.

35 Ibid., 31

36 See "Building the Hudson River Vehicle Tunnel," *Engineering News-Record* 92 (May 8, 1924): 798–803.

37 WJW, "Preliminary Estimates of 1919 and Estimated Cost of Completed Tunnel under Current Plans," Dec. 20, 1923, WWP, box 49.

38 "Largest American Shield Tunnel Designed to Carry Vehicular Traffic under Hudson River," *Engineering News-Record* 84 (Feb. 19, 1920): 357.

39 Ibid., 361.

40 Robert Skerrett, "The World's Greatest Vehicular Tunnel," *Scientific American*, May 8, 1920, 510.

41 George Dyer to WJW, May 4, 1922, WWP, box 49, Correspondence.

42 "Gas Bomb Fumes Test Holland Tunnel," *NYT*, Mar. 16, 1927, 2.

43 "Seek Bus Rights in Vehicular Tunnel," *NYT*, Feb. 2, 1927, 3.

44 "Jersey Road Link Will Open July 4," *NYT*, June 19, 1927, E21.

45 "Fears Truck Jam in Holland Tube," *NYT*, Sept. 17, 1927, 21.

46 "The Holland Tunnel Is a Modern Marvel," *NYT*, Oct. 9, 1927.

47 "Rules Are Changed in Holland Tunnel," *NYT*, Nov. 22, 1927, 13.

48 "Holland Tunnel Rounds Out a Successful Year," *NYT*, Oct. 11, 1928, 159.

49 Jameson Doig, *Empire on the Hudson: Entrepreneurial Vision and Political Power at the Port of New York Authority* (New York: Columbia University Press, 2001), 89.

50 "Tunnels and the Port of New York," *Engineering News-Record* 84 (Feb. 21, 1924): 309.

51 "Address by William J. Wilgus at the Regular Meeting of the New York Section of the American Society of Civil Engineers," Feb. 13, 1925, WWP, box 49.

52 See Doig, *Empire on the Hudson,* 169–71.

53 WJW, "Milestones," 212.

54 Ibid., 223.

55 Ole Singstad to WJW, Nov. 29, 1927, WWP, box 49, Correspondence.

56 WJW to Ole Singstad, Aug. 15, 1930, WWP, box 49, Correspondence.

Chapter 7. Joining Staten Island to New York City

1 "Names $324,478,300 to Reproduce Road," *NYT*, Apr. 23, 1919, 10.

2 "Belt Line Plan Recommended for Port of New York," *Engineering News-Record* 86 (Feb. 10, 1921): 271–72.

3 New York State, Laws of 1921, chap. 700.

4 WJW, "The Cost of Terminal Freight Handling in the Port of New York," *Engineering News* 63 (Mar. 31, 1910): 360–61.

5 WJW, "Milestones," 213.

6 "City Engineers Back Narrows Tunnel Plan," *NYT*, Dec. 29, 1921, 5.

7 WJW, "New York's Proposed Belt Railway," *Scientific American*, Jan. 1922, 40.

8 WJW, "Milestones," 214.

9 "City Tunnel Cost Put at $110,000,000," *NYT*, Jan. 26, 1922, 10.

10 WJW to E. B. Temple, Assistant Chief Engineer, Pennsylvania Railroad, Dec. 6, 1921, WWP, box 55, Narrows Tunnel & Outerbridge.

11 "City's Future on Port Fight Plan," *NYT*, Feb. 5, 1922, 82.

12 John J. Rooney, "Hylan Side of Port Controversy," *NYT*, Feb. 19, 1922, 94.

13 Robert Caro, *The Power Broker: Robert Moses and the Fall of New York* (New York: Vintage, 1974).

14 WJW to Arthur Tuttle, Chief Engineer, Board of Estimate & Appropriation, Jan. 18, 1922, WWP, box 53, folder 4.2.

15 Ibid., 11.

16 Ibid., 22.

17 Ibid., 34.

18 "Interchange via Bay Ridge: Freight from New Jersey to Long Island Railroad Freight Yard in Bay Ridge, Brooklyn, for Distribution via Long Island Railroad," Table A, Tonnages and Cars, 1921, WWP, box 54, folder 13–Tonnages.

19 WJW, "Statement at Joint Hearing Given by the Committee on Finance of the Senate and Committee on Ways and Means of the Assembly on January 31, 1922," 1, WWP, box 54, folder 8–Brooklyn–Richmond Tunnel.

20 Ibid., 6.

21 WJW, "Milestones," 216.

22 Ibid.

23 "Hylan Swings Pick at Shaft Opening," *NYT*, Jan. 20, 1923, 28.

24 "The Narrows Tunnel," *NYT*, May 3, 1922, 17.

25 WJW, "Milestones," 217.

26 "Hylan Tunnel Plan Doomed in Albany," *NYT*, Mar. 4, 1923, 8.

27 "Smith Is Expected to Rebuff Hylan and Sign Tube Bill," *NYT*, Apr. 21, 1925, 1.

28 Port Authority of New York and New Jersey, *Annual Report, 1926,* 5.

29 Jameson Doig, *Empire on the Hudson: Entrepreneurial Vision and Political Power at the Port of New York Authority* (New York: Columbia University Press, 2001), 114, 119.

30 "Regional Transportation Statistical Report," New York Metropolitan Transportation Council, Mar. 2009, 58, table B-7.

31 Edwards & Kelcey Engineers, "Cross Harbor Freight Movement Investment Study," May 2000, www.crossharborstudy.com.

32 See Henry Petroski, *Engineers of Dreams* (New York: Knopf, 1995), 237–72.

33 Thomas Adams, Director of Plans, Regional Plan of New York and Its Environs, to WJW, Feb. 26, 1926, WWP, box 56, folder 55.4-Regional Plan of New York Fort Washington Bridge.

34 WJW to Thomas Adams, Mar. 3, 1926, WWP, box 56.

35 WJW to Adams, Sept. 4, 1926, WWP, box 56.

36 Adams to WJW, Sept. 9, 1926, WWP, box 56.

37 O. H. Amman to Thomas Adams, Sept. 21, 1926, WWP, box 56.

38 "Final Hearing Backs Hudson Bridge Plan," *NYT*, Dec. 3, 1926, 25.

39 WJW to Thomas Adams, Dec. 26, 1926, WWP, box 56.

40 Colonel Frederic Delano, Chairman, Regional Plan of New York and Its Environs, to George Silzer, Chairman of the Port of New York Authority, Dec. 15, 1926, WWP, box 56.

41 Ole Sinstad, Chief Consulting Engineer on Tunnels, Port of New York Authority, to WJW, July 17, 1930, WWP, box 56.

42 Arthur Warner, "An Astounding Span of Steel and Wire," *NYT*, Oct. 18, 1931, 119.

43 "Washington Bridge a Year Old," *NYT*, Oct. 21, 1932, 21.

44 "Tunnels and Spans Weather the Slump," *NYT*, Oct. 31, 1932, 31.

45 "Income Rise Shown by Port Authority," *NYT*, Mar. 9, 1937, 46.

46 Charles Bagli and N. Confessore, "Take the No. 7 to Secaucus? That's a Plan," *NYT*, Nov. 16, 2010, A1.

47 WJW, "Milestones," 219.

48 "Delamater Quits as Works Director; Colonel William J. Wilgus to Succeed Him in Post in Welfare Department," *NYT*, Aug. 13, 1934, 15.

49 WJW, "Milestones," 245.

50 "95 Bosses, 91 Men Are Found Working on City Relief Job," *NYT*, Dec. 7, 1934, 1.

51 "Hodson Denounces Relief Testimony," *NYT*, Dec. 9, 1934, 38.

52 WJW, "Milestones," 246.

53 Ibid., 269.

54 Ibid., 254.

55 Ibid., 258.

Conclusion

1 "Secret Wedding Revealed," *NYT*, Nov. 3, 1911, 11.

2 "W. J. Wilgus Jr. Better after Dose of Poison," *NYT*, Jan. 10, 1927, 25.

3 William J. Wilgus Jr. to John Oppenheim, July 20, 1931, James Oppenheim Papers, 1898–1932, Manuscripts Division, New York Public Library, box 2.

4 Wilgus Jr. to Oppenheim, Nov. 11, 1931.

5 Wilgus Jr. to Oppenheim, Nov. 22, 1931, 1.

6 Ibid., 2.

7 Wilgus Jr. to Oppenheim, April 4, 1932.

8 John Hurd, *Weathersfield, Century Two: The Story of a Small Vermont Town during Its Second Hundred Years* (Weathersfield, VT: Weathersfield Historical Society, 1978), 124.

9 State of New Hampshire, Sullivan County, Court of Probate, "Inventory of, and Apprise the Estate of William J. Wilgus, Nov. 18, 1949."

10 WJW, "Milestones," 339.

11 Ibid., 344.

12 Ibid., 351–56.

13 Ibid., 359.

14 New York Metropolitan Transportation Council, *2005–2030 Regional Transportation Plan*, August 2005, 20.

15 "State Auto Fees Amount to $856,310," *NYT*, June 14, 1912, 9.

16 "Auto Census in New York," *NYT*, Jan. 9, 1921, 82.

17 Robert Caro, *The Power Broker: Robert Moses and the Fall of New York* (New York: Knopf, 1974).

18 "William J. Wilgus, Rail Expert, Dead," *NYT*, Oct. 25, 1949, 27.

Index

Page numbers in italics indicate figures, maps, and tables.

New York Central Railroad
(*continued*)
76–78, 81–82, 85, 91–92, 96,
104–5; corporate offices of, 42,
43; freight subway proposal
and, 94, 95, 100; High Line
used exclusively by, 105, *136*,
136–37, 139; Inter-Terminal
Belt Line proposal and, 119, 120,
124–25, 126, 130, 131–32; law-
suit threatened by, 91–92; New
Jersey shipping compared with,
169–70; nosing problem cover-
up of, 65; origins of, 258n21; rail
yards at 30th and 60th Streets,
79, 79–85, *80*; revenues of
(1901–2), 48–49; truck shipping's impact on, 139–40; Vanderbilts' control of, 25–26;
WJW's advancements and positions with, 6, 42–43, 63; WJW's
father's work for, 3; WJW's resignation from, 63, 66, 73. *See
also* Grand Central Terminal;
New York Central & Hudson
River Railroad
New York City: constitutional
debt limit of, 113; electrification, 50–51, 57; Erie Canal's
importance to, 18–20; freight
train tracks on streets of, 25,
26, 43, 74–75, *76*, 76–78, 81–
82, 85, 91–92, 96, 104–5; geography, 11–14, *12*; immigration,
26–28, *27*, *28*; legal battles
against New York Central, 132,
133–34, 135–36, 170; manufacturing, 7–8, 29–31, *30*; name
of, 15; political tensions with
Port Authority, 219–20, 226;
politics and corruption in, 95–
97, 101–2, 113; protest march
in (1915), 151; railroads drawn
to, 36–41; rise to dominance, 7–
8, 15–23, 31, *75*, 165, 173; transportation challenges in, 8–10,
31–36, *32*, *33*; transportation
challenges ongoing in, 239–42,
251–56; WJW's arrival in, 6, 11,
31, 41; WJW's Grand Central
proposal in context of, 48–49;

WWI and, 162–64. *See also*
port of New York
New York City & Northern Rail-
road, 125–26
New York City Board of Estimate,
220, 222, 226–30, 231, 232–33
New York Connecting Railroad,
38, 146, 148, 192, 220–21
New York Harbor. *See* port of New
York
New York Harbor case, 150, 164,
165
New York Metropolitan Transpor-
tation Council, 234–35, 252
New York, New Haven & Hartford
Railroad, 219
New York–New Jersey Bridge and
Tunnel Commission (earlier,
New York Interstate Bridge
Commission): appreciation for
WJW, 218; on consulting engineers' role, 196–97; proposed
bridge and, 193; public hearings of, 199; study of, 191; ventilation system and, 210. *See
also* Board of Consulting Engineers (Holland Tunnel)
New York, New Jersey Port and
Harbor Development Commission: automatic-electric system
plan of, 174–77, *175*, *176*, *177*;
basic premise of, 165–66; debate about, 149–51; establishment of, 33, 103–4, 164; Metropolitan District idea and, 144,
163; Port Authority recommended by, 177–79; reports of,
33, 34–35, 78, 164–74
New York Public Improvement
Committee, 236
New York Public Library, WJW
materials in, 154, 163, 245, 251
New York State Board of Railroad
Commissioners, 46, 64
New York State Legislature: elevated railroad charter, 110–11;
franchise terms, 117; George
Washington Bridge bill, 236;
highway commission estab-
lished, 254; Holland Tunnel
bill, 195–96; Holland Tunnel

turned over to Port Authority,
216–17; immigrant arrival and
registration, 27; Manhattan
freight situation studied, 103–
4; Narrows Tunnel project and,
220, 227, 231–33; railroad
mergers and, 78; rapid transit
issues, 113, 132–33; right to use
streets for train tracks and, 76–
77, 133; steam vs. electric
power issue, 46, 53; street railway charters, 108–9; "west side
problem" and, 81–82
New York State Transit Commission, 162
North German Lloyd Line, 32,
118, 173
North River Bridge Company,
192–93
nosing problem, 65

Ogdensburg & Lake Champlain
Railroad, 42
Oldsmobile Company, 9
omnibuses, 106, 108, 109, 110, 139
operating ratio concept, 111
Oppenheim, James, 246–47
O'Rourke Construction Company,
60, 62
Outerbridge, Eugene, 151

packets. *See* steamships and
steamboats
Park Avenue tunnel and crash,
44–46, *45*
Parsons, William, 85
passenger service: competition in,
46–47; decline due to automobile age, 239–41; freight shipping revenues vs., 36, 38, 48,
241; internal circulation issue
of, 106–7; maintained during
construction, 61; numbers and
congestion in, 43–47, 115–16,
117–18, 128–30; omnibuses as,
106, 108, 109, 110, 139; renewed plan for, 241–42, 255;
terminals for, 25–26, *68*; transatlantic passage, 22–23, 26–28,
27, *28*; vehicles as competition
for railroads, 213–14; world-

Illustration Credits

Atlas of the Borough of Manhattan, City of New York (New York: G. W. Bromley, 1902), sec. 2, plate 39: p. 85

Atlas of the Borough of Manhattan, City of New York (New York: G. W. Bromley, 1916), sec. 4, plate 2: p. 79; sec. 3, plate 16: p. 80 (A); sec. 3, plate 17: p. 80 (B)

Grand Central Terminal under Construction, ca. 1909, Museum of the City of New York, Print Archives: p. 67

Friends of the High Line website, www.thehighline .org: p. 136

New York Daily News via Getty Images: p. 215

New York Public Library, Astor, Lenox and Tilden Foundations:

Photography Collection, Miriam and Ira D. Wallach Division of Art, Prints and Photographs: pp. 35, 109

Milstein Division of United States History, Local History & Genealogy: pp. 138, 188

Science, Industry & Business Library: pp. 204, 205, 210, 211

Picture Collection: pp. 21, 26, 32, 114, 187, 189

William J. Wilgus Papers, Manuscripts Division, New York Public Library: frontispiece and pp. 60, 61, 156, 159

Adapted from Wilgus, "Proposed New Railway System for the Transportation and Distribution of Freight by Improved Methods in the City and Port of New York, 1908," box 45: p. 75

Maps by Bill Nelson: pp. 12, 19, 45, 71, 87, 91, 107, 123, 147, 154, 157, 164, 175, 181, 221, 225, 253